Flame Spray Drying

Advances in Drying Science and Technology

Series Editor
Arun S. Mujumdar
McGill University, Quebec, Canada

Thermal and Nonthermal Encapsulation Methods
Magdalini Krokida

Industrial Heat Pump-Assisted Wood Drying
Vasile Minea

Intelligent Control in Drying
Alex Martynenko and Andreas Bück

**Drying of Biomass, Biosolids, and Coal: For Efficient Energy Supply
and Environmental Benefits**
Shusheng Pang, Sankar Bhattacharya, Junjie Yan

Drying and Roasting of Cocoa and Coffee
Ching Lik Hii and Flavio Meira Borem

Heat and Mass Transfer in Drying of Porous Media
Peng Xu, Agus P. Sasmito, and Arun S. Mujumdar

Freeze Drying of Pharmaceutical Products
Davide Fissore, Roberto Pisano, and Antonello Barresi

Frontiers in Spray Drying
Nan Fu, Jie Xiao, Meng Wai Woo, Xiao Dong Chen

Drying in the Dairy Industry
*Cécile Le Floch-Fouere, Pierre Schuck, Gaëlle Tanguy, Luca Lanotte,
Romain Jeantet*

Spray Drying Encapsulation of Bioactive Materials
Seid Mahdi Jafari and Ali Rashidinejad

Flame Spray Drying: Equipment, Mechanism, and Perspectives
Mariia Sobulska and Ireneusz Zbicinski

For more information about this series, please visit: www.crcpress.com/Advances-in-Drying-Science-and-Technology/book-series/CRCADVSCITEC

Flame Spray Drying
Equipment, Mechanism, and Perspectives

Mariia Sobulska and Ireneusz Zbicinski

CRC Press
Taylor & Francis Group
Boca Raton London New York

CRC Press is an imprint of the
Taylor & Francis Group, an **informa** business

First edition published 2022
by CRC Press
6000 Broken Sound Parkway NW, Suite 300, Boca Raton, FL 33487-2742

and by CRC Press
2 Park Square, Milton Park, Abingdon, Oxon, OX14 4RN

© 2022 Taylor & Francis Group, LLC

CRC Press is an imprint of Taylor & Francis Group, LLC

ISBN: 978-0-367-56931-0 (hbk)
ISBN: 978-0-367-60762-3 (pbk)
ISBN: 978-1-003-10038-6 (ebk)

DOI: 10.1201/9781003100386

Typeset in Times
by codeMantra

Contents

Preface

Drying processes being one the most energy-consuming operations in the industry require innovations in terms of improvement of energy efficiency and reduction of environmental impact. This book provides recent advances in the field of novel flame spray drying (FSD) and similar flame drying techniques focusing on the mechanism, equipment, potential applications, and the effect of operating parameters on the dry product quality, energy consumption, and safety issues.

The book is divided into four chapters. The first chapter presents the mechanism of FSD based on the measurements of dispersed and continuous-phase parameters (PSD, particle velocities, air and particle temperatures, air flow pattern, residence time, powder properties, etc.). The chapter also includes FSD CFD model with modifications of the Navier–Stokes equations reflecting combustion reactions of the species allowed to determine basic mechanism of FSD process and relationship between particle residence time in flame and drying time to control quality of the dried product. The description of FSD process mechanism is followed by recommendations and suggestions for FSD scaling-up procedure including basic technical and safety requirements.

The second chapter presents the examples of FSD process application for dewatering of thermally resistant (ceramic material) and thermally sensitive (maltodextrin and coffee) materials. The results of the experimental measurements of dried powder properties including bulk and apparent densities, particles morphology, moisture and fuel content, color index, and hydroxymethylfurfural content are shown and discussed. The effect of FSD parameters such as solid and fuel content in the feed, feed rate, atomization pressure, air flow rate in the tower, and feed temperature on particle size distribution and product quality is described.

The third chapter presents the principles, mechanism, equipment, and applications of three flame techniques: flame drying of textiles (Remaflam® process), pulse combustion drying, and flame spray pyrolysis. Application of flame for drying and particles synthesis provides numerous advantages such as process intensification, reduction of drying time, possibility to provide the process in single step, and enhanced energy efficiency.

The fourth chapter covers analysis of FSD process safety operation, energy consumption, and environmental issues, followed by summary of the main advantages and disadvantages of FSD technique and perspectives of further development. Energy consumption of FSD process is evaluated and compared with conventional spray drying method proving lower energy consumption in FSD.

This book, with theoretical background of drying and combustion processes as well as practical examples of flame drying applications, can be useful for engineers, researchers, and students in the field of drying technology, and chemical and process engineering.

Acknowledgments

The authors are grateful to Dr. Marcin Piatkowski for his irreplaceable help with preparation of experimental installation and participation in experimental part, assistance with microscopic and image analysis, his valuable comments in interpreting the results of the experiments, and for sharing his knowledge and expertise during all research work.

We would like to thank the company Ceramika Paradyz Inc. (Opoczno, Poland) for the supply of ceramic material and company Nestle Poland S.A. (Poland) for supply of soluble coffee applied in the research.

Advances in Drying Science and Technology

SERIES EDITOR: DR. ARUN S. MUJUMDAR

It is well known that the unit operation of drying is a highly energy-intensive operation encountered in diverse industrial sectors ranging from agricultural processing, ceramics, chemicals, minerals processing, pulp and paper, pharmaceuticals, coal polymer, food, forest products industries as well as waste management. Drying also determines the quality of the final dried products. The need to make drying technologies sustainable and cost effective via application of modern scientific techniques is the goal of academic as well as industrial R&D activities around the world.

Drying is a truly multi- and interdisciplinary area. Over the last four decades the scientific and technical literature on drying has seen exponential growth. The continuously rising interest in this field is also evident from the success of numerous international conferences devoted to drying science and technology.

The establishment of this new series of books entitled Advances in Drying Science and Technology is designed to provide authoritative and critical reviews and monographs focusing on current developments as well as future needs. It is expected that books in this series will be valuable to academic researchers as well as industry personnel involved in any aspect of drying and dewatering.

The series will also encompass themes and topics closely associated with drying operations, e.g., mechanical dewatering, energy savings in drying, environmental aspects, life cycle analysis, technoeconomics of drying, electrotechnologies, control and safety aspects, and so on.

ABOUT THE SERIES EDITOR

Dr. Arun S. Mujumdar is an internationally acclaimed expert in drying science and technologies. He is the Founding Chair in 1978 of the International Drying Symposium (IDS) series and Editor-in-Chief of Drying Technology: An International Journal since 1988. The fourth enhanced edition of his *Handbook of Industrial Drying* published by CRC Press has just appeared. He is the recipient of numerous international awards including honorary doctorates from Lodz Technical University, Poland, and University of Lyon, France.

Please visit www.arunmujumdar.com for further details.

Authors

Dr. Mariia Sobulska is an assistant professor in the Department of Environmental Engineering at the Faculty of Process and Environmental Engineering in the Lodz University of Technology, Poland. She obtained a PhD in Chemical Engineering from Lodz University of Technology in 2019. Her research interest is in spray drying process and fabrication of food powders, heat and mass transfer, mathematical modeling, and energy saving in chemical processes. In 2015, she was awarded at Nordic Baltic Drying Conference held in Gdansk, Poland, "for distinguished contribution to the development of drying technology and dissemination to Nordic Baltic Region." She was a participant in the project devoted to drying of graphene funded by Grupa Azoty S.A. (Tarnów, Poland) and the project on development of chemical heat pump funded by Polish National Centre for Research and Development.

Professor Ireneusz Zbicinski completed a PhD at the Faculty of Chemical Engineering in 1981. Since 2001, he has worked as a full professor at Lodz University of Technology. His scientific and industrial training covered Chemical Plants "Anilana" Lodz, 1982, EA Technology, Capenhurst, UK, 1999–2000, University of Birmingham, UK, 1989, Washington State University, Pullman, USA, 1996. His research activity covers areas of chemical engineering processes, drying, spray drying, LCA. He is the holder of Qilu Friendship Award granted by Government of Shandong Province, China, for contribution to social and economic development of the province, 2010, and the "Excellence in Drying Research" Award from an international committee of scientists and industry representatives. He took part in more than 25 international and domestic scientific projects, e.g., for Procter&Gamble, Cabbot, International Fine Particle Research Institute, UK/USA, Henkel, Anwil, Chemat, etc. He lectures at universities and industrial labs in USA, France, Germany, Holland, Sweden, Japan, China, Brazil, etc. He is a member of steering committee of EU INTERREG IIIB project Watersketch (2004–2007) and Waterpraxis (2008–2011), INVITES (2016-), etc. He is involved in three Horizon 2020 projects on LCA on TCO (INREP, 2014–2017), on CPV (Hiperion, 2020–2023), and on light boats (Fiber4Yards, 2021–2023). He supervised 16 PhD students and has published over 70 papers in top-ranked scientific journals (h-index = 19).

1 Principles and Mechanism of Flame Spray Drying

1.1 PRINCIPLES OF FLAME SPRAY DRYING PROCESS

Drying as one the most energy-consuming operation is responsible for about 10%–25% of energy consumption in the industrial sector (Law and Mujumdar 2010). Spray drying is a widely applied technique for dewatering of the liquid feedstock such as solutions, slurries, emulsions, and pastes into powder product. Due to low thermal efficiency (Filková, Huang, and Mujumdar 2014), spray drying process consumes large amounts of energy (4,500–11,500 kJ/kg H_2O) compared with other drying methods such as tunnel dryers (5,500–6,000 kJ/kg H_2O) or fluid bed dryers (4,000–6,000 kJ/kg H_2O) (Zbicinski 2002).

The scheme of standard spray drying (SSD) process is shown in Figure 1.1. In the standard spray dryer, the air is typically heated up in the direct or indirect heaters, and the liquid feed is dispersed into small droplets dried in the hot drying medium. In the direct heaters, the flue gases from combustion of coal, gas, or oil are mixed with air to decrease the gas temperature and supplied to the spray dryer. In the indirect heaters, the drying air is heated up by the steam, electric energy, or flue gases without contact with heating medium. Application of direct-fire heating systems with natural gas combustion increases the energy efficiency of drying process compared with steam-heated and other types of indirectly heated dryers (Strumillo, Jones, and Zylla 2006). Application of electric heating is limited only to small-scale dryers due to the high cost of electric energy, which is three times higher than that of conventional fossil fuels (Kemp 2011). The hot drying air is supplied to the spray dryer by fan, and the liquid feed is supplied by the pump to the nozzle installed inside the drying chamber. After spray drying, the dry product in the powder form is separated from the air stream in the cyclones, bad filters, or scrubbers.

According to Strumillo, Jones, and Zylla (2006), energy-saving measures, which can be applied in the convective drying processes, among others, include the following:

- Utilization of biomass or biofuels as energy sources
- Reduction of the drying time and dryer length to decrease the area of the dryer, where heat loss to the environment may occur
- Supply of the energy exactly where it is used, i.e., to the drying chamber to avoid the heat losses in the air heater
- Reduction of drying air consumption

DOI: 10.1201/9781003100386-1

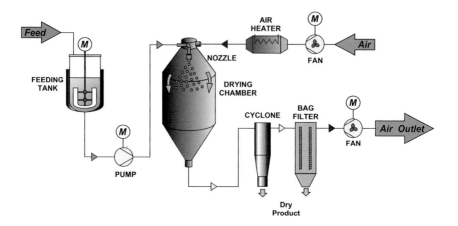

FIGURE 1.1   Scheme of standard spray drying process.

- Increase of the driving force of the process, i.e., increase of the inlet temperature of the drying medium
- Application of intermittent drying or pulsation flow of the drying medium

Generation of energy precisely in the place where it is needed, i.e., directly in the zone of moisture evaporation to avoid energy losses to the environment in the auxiliary equipment, has already been utilized in the flame drying of textiles (Remaflam® process; Eltz et al. 1985) and can also be applied in the disperse systems.

One of such an opening is flame spray drying (FSD), a novel spray drying method utilizing combustion of flammable component of the spray as an energy source for drying process, which has been developed and patented at the Faculty of Process and Environmental Engineering, Lodz University of Technology (Piatkowski and Zbicinski 2013). In the FSD process, energy required to evaporate a solvent comes from the combustion of flammable spray component, which makes drying installation independent from conventional energy sources such as gas, oil, or electricity. Moreover, FSD process gives possibility to apply different types of liquid biofuels, i.e., bioethanol or vegetable oils coming from renewable energy sources for drying process, and decrease the emission of harmful gases to the atmosphere.

FSD consists of the following steps: mixing of flammable component with raw material, ignition, and fuel combustion, which generates heat used for moisture evaporation and particles drying (Figure 1.2). There is no air heating system in FSD process, which helps to reduce investment costs and heat losses from auxiliary equipment.

1.2   TECHNICAL REQUIREMENTS IN FLAME DRYING

1.2.1   CONSTRUCTION MATERIALS

The main requirement for drying chamber construction material in flame drying is resistance to high-temperature operations for a long period of time. The additional specifications for construction materials come from mechanical stresses generated by temperature fluctuations within the flame spray dryer.

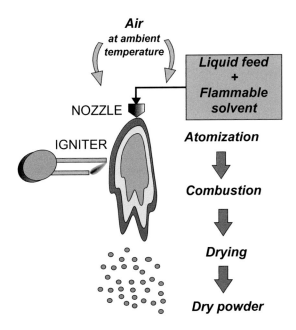

FIGURE 1.2 Principle of flame spray drying process.

The materials applied in the construction of FSD drying chambers should have the following characteristics:

- High thermal resistance (Kudra 2008)
- High mechanical strength
- High resistance to oxidation and corrosion (Romero-Jabalquinto et al. 2016)
- Low thermal expansion
- High resistance to thermal fatigue
- Low costs
- High availability
- Ease to use in manufacture process

The list of materials resistant to high temperatures with high strength and toughness commonly applied for construction of industrial furnaces and combustion chambers includes stainless steel, iron-based alloys, cobalt-based superalloys, and nickel-based superalloys (Zonfrillo, Giovannetti, and Manetti 2008). Stainless steels with high-temperature properties include both ferritic (e.g., AISI 446) and austenitic type of steel (e.g., AISI 314, AISI 310). Compared with austenitic steel, ferritic steels have lower yield strengths, whereas superalloys have the highest strength within the discussed materials. Stainless steel is a less expensive material than superalloys, the cost of which is increasing in the following order: iron-based alloys, nickel-based superalloys, and cobalt-based superalloys (Yang et al. 2002).

Within each material group, selected alloys are compared in terms of composition, costs, high-temperature resistance, oxidation resistance, and manufacturability

TABLE 1.1

Comparison of Selected Materials for High-Temperature Applications

Material	Composition	Cost	Characteristic	Reference
			Stainless Steel	
AISI 310	0.25C, 25Cr, 21Ni, 2Mn, 1.5Si	Low	Heat-resistant up to 1,150°C, creep-resistant up to 690°C, medium mechanical strength, good oxidation resistance, good manufacturability	Zonfrillo, Giovannetti, and Manetti (2008), Yang et al. (2002)
AISI 316	0, 25C, 25Cr, 21Ni, 2Mn, 2.5Si	Low	Heat-resistant up to 1,150°C, creep-resistant up to 690°C, medium mechanical strength, good oxidation resistance, good manufacturability	Zonfrillo, Giovannetti, and Manetti (2008), Yang et al. (2002)
AISI 446	0, 2C, 26Cr, 1.5Mn, 1Si	Low	Resistant to temperatures up to 1,100°C, low mechanical strength, good oxidation resistance, medium manufacturability	Zonfrillo, Giovannetti, and Manetti (2008), Yang et al. (2002)
			Iron-Based Alloys	
HR-120	37Ni, 25Cr, 3Co, 2.5Mo, 2.5W, 0.7Nb, 0.7Mn, 0.6Si	Medium	Improved creep rupture strength, high mechanical strength, good oxidation resistance, good manufacturability	Zonfrillo, Giovannetti, and Manetti (2008), Yang et al. (2002)
Multimet®	21Cr, 20Ni, 20Co, 3Mo, 2.5W, 1.5Mn, 1Nb+ Ta, 1Si	Medium	May be applied at high mechanical stress at temperatures ca. 800°C and medium mechanical stress at temperatures up to 1,000°C; good oxidation resistance, good manufacturability	Zonfrillo, Giovannetti, and Manetti (2008), Yang et al. (2002)
			Cobalt-Based Superalloys	
Haynes® 188	22Ni, 22Cr, 14W, 3Fe, 1.25Mn, 0.35Si, 0.03La	Very high	Very high strength and oxidation resistance at temperatures up to 1,000°C, medium manufacturability	Zonfrillo, Giovannetti, and Manetti (2008), Yang et al. (2002)
Haynes® 25	20Cr, 10Ni, 15W, 3Fe, 1.5Mn, 0.4Si	Very high	Very high strength and oxidation resistance at temperatures up to 980°C, medium manufacturability	Zonfrillo, Giovannetti, and Manetti (2008), Yang et al. (2002)
			Nickel-Based Superalloys	
Nimonic® 263	20Cr, 20Co, 6Mo, 2Ti, 0.7Fe, 0.6Al, 0.6Mn, 0.4Si, 0.2Cu	High	Very high strength and oxidation resistance at temperatures up to 750°C, good manufacturability	Zonfrillo, Giovannetti, and Manetti (2008), Yang et al. (2002), "Nimonic 263 Nickel Based Alloy Supplier I HARALD PIHL I Harald Pihl" (2020)

(Continued)

TABLE 1.1 (*Continued*)
Comparison of Selected Materials for High-Temperature Applications

Material	Composition	Cost	Characteristic	Reference
Hastelloy®-S	16Cr, 15Mo, 3Fe, 2Co, 1W, 0.5Si, 0.4Al, 0.3Cu, 0.05La	High	Very high strength and oxidation resistance at temperatures up to 1,093°C, good manufacturability	Zonfrillo, Giovannetti, and Manetti (2008), Yang et al. (2002), "Haynes International - Principal Features Hastelloy S" (2020)
Hastelloy® -X	22Cr, 19Fe, 9Mo, 1.5Co, 1Si, 1Mn, 0.7W	High	Very high strength and oxidation resistance, resistant to stress-corrosion cracking, good manufacturability	Zonfrillo, Giovannetti, and Manetti (2008), Yang et al. (2002), "Haynes International - Principal Features Hastelloy X" (2020)
Inconel® 740	25Cr, 20Co, 2Nb, 1.8Ti, 0.9Al, 0.7Fe, 0.5Mo, 0.5Si	Very High	Very high strength and oxidation resistance, good manufacturability	Zonfrillo, Giovannetti, and Manetti (2008), Yang et al. (2002)

(Table 1.1). Ferritic chromium stainless steel (AISI 446) is resistant to temperature up to 1,100°C; however, at the temperature above 700°C, its creep and fatigue strength decrease significantly. Additional drawback of ferritic steels as construction material is lower manufacturability compared with austenitic steels and superalloys. Austenitic chromium–nickel–silicon steels such as AISI 314 and AISI 310 are resistant to high temperature of 1,150°C and are creep-resistant at temperatures up to 690°C.

Among the superalloys, the following materials are typically used in the construction of combustion chambers: iron-based alloys (HR-120, Multimet®), nickel-based superalloys (Nimonic® 263, Hastelloy®-S, Hastelloy®-X, Inconel® 740), and cobalt-based superalloys (Haynes® 188, Haynes® 25).

In order to reduce volumetric expansion, and thermal and mechanical stress, the drying chamber design may be also modified.

1.2.2 Temperature Control in Flame Drying

Accurate measurement of temperature within the combustion zone enables precise control of the process in terms of product quality, efficient energy utilization, and reduction of environmental pollutions, and plays a key role in safe operation of FSD. For temperature measurements within the flame, two different methods may be applied, which include invasive and noninvasive techniques.

1.2.2.1 Invasive Techniques of Temperature Measurements in Flame

Invasive techniques are based on the direct insertion of a probe or sensor within the zone where desired parameters should be measured. For combustion systems, which are characterized by the high temperatures and high amplitude of temperature fluctuations, the following invasive techniques of temperature measurements may be applied: thermocouples, resistance thermometry, and thermochrome paintings (Leipertz, Pfadler, and Schießl 2010).

Thermocouples are widely used for temperature measurements in both industrial combustion systems and research installations. The basic principle of temperature measurement via thermocouple may be described as follows: electric voltage is generated when temperature difference occurs between two metals with different conductivity joined into the closed loop.

For temperature control within the combustion chamber of flame spray dryer, application of thermocouples resistant to hot temperatures, which are typically used in industrial furnaces, is recommended. The following types of thermocouples can be applied in the high-temperature environment:

- "type J" thermocouple is composed of positive iron wire and negative copper–nickel alloy wire (about 45% of nickel and 55% of copper); "type J" thermocouple is applicable in the range of temperatures from −180°C to 800°C; the measurement accuracy of "type J" thermocouple of class 1 varies depending on the temperature range: for lower temperatures from −40 to +375, accuracy is ±1.5°C, and for higher temperatures from +375°C to +750°C, accuracy is ±0.4% ("International Standard IEC 60584-1:2013. Thermocouples – Part 1: EMF Specifications and Tolerances" 2013).
- "type K" thermocouple is composed of nickel–chromium wire (90% nickel and 10% chromium) and nickel–aluminum wire (95% nickel, 2% manganese, 2% aluminum, and 1% silicon); "type K" thermocouple can be used in the range of temperatures from −180°C to 1,200°C; the measurement accuracy: for temperatures in the range from −40°C to +375°C, accuracy is ±1.5°C, and for higher temperatures from +375°C to +1,000°C, accuracy is ±0.4% ("International Standard IEC 60584-1:2013. Thermocouples – Part 1: EMF Specifications and Tolerances" 2013).
- "type S" thermocouple is composed of positive platinum–rhodium wire (about 90% platinum, 10% rhodium) and negative platinum wire, which makes this thermocouple type applicable in the range of temperatures from −50°C to 1,750°C; the measurement accuracy: for temperatures in the range from 0°C to +1,100°C, accuracy is ±1°C, and for higher temperatures from +1,100°C to +1,600°C, accuracy is described by the expression ±(1°C + 0.003*(t°C − 1100°C))°C ("International Standard IEC 60584-1:2013. Thermocouples – Part 1: EMF Specifications and Tolerances" 2013).
- "type B" thermocouple is composed of positive platinum–rhodium wire (about 70% platinum, 30% rhodium) and negative platinum–rhodium wire

(about 94% platinum, 6% rhodium), which makes this thermocouple type applicable in the range of temperatures from 0°C to +1,820°C; the measurement accuracy is about ±0.25% ("International Standard IEC 60584-1:2013. Thermocouples – Part 1: EMF Specifications and Tolerances" 2013).

- "type N" thermocouple is a nickel/chromium (14.2%)/silicon (1.4%) and nickel/silicon (4.4%) thermocouple capable of operating at the temperature range from −270°C to 1,300°C; the measurement accuracy: for temperatures in the range from −40°C to +375°C, accuracy is ±1.5°C, and for higher temperatures from +375°C to +1,000°C, accuracy is ±0.4% ("International Standard IEC 60584-1:2013. Thermocouples – Part 1: EMF Specifications and Tolerances" 2013).

Additionally, the thermocouples should be covered with protection tubes constructed from one of the following materials: stainless steel AISI 446, AISI 314, or other heat-resistant materials such as Hastelloy®, Kanthal®, Alloy 20, Inconel® 600, and ceramic materials (JumoIndustries 2014).

The main advantages of temperature measurements within flame via thermocouples are low cost, high accuracy, and easy and simple operation (Heitor and Moreira 1993). The application of thermocouples for measurements in flame as an invasive technique has also disadvantages: intervention into the flow dynamics, local mixing rate of fuel and oxidizer, measurement errors caused by contamination of thermocouple wires, catalytic effect due to chemical reaction of thermocouple material with chemically reactive species, and errors arising from radiative and conductive heat exchange between the thermocouple and flame (Heitor and Moreira 1993).

Another invasive technique used for temperature measurements in flame is resistance thermometry (Venkataraman et al. 2020). The principle of resistance thermometer is based on the measurement of electrical resistance of conductor material, which depends on the temperature changes. In resistant thermometers, the following materials could be applied: nickel, copper, platinum (i.e., Pt100), or semiconductor materials (Hahn and Edgar 2003). The disadvantages of utilization of resistance thermometers in flame measurements are narrow range of temperatures (which does not exceed 850°C) and long response time (Leipertz, Pfadler, and Schießl 2010). The advantage of resistance thermometry is higher precision of measurements compared with thermocouples.

Recent studies describe the application of thermochromic paintings for measurements of surface temperature of combustion chambers in the laboratory scale (Arulprakasajothi and Rupesh 2020) and industrial (Sallevelt et al. 2014) gas turbine combustors. The main principle of temperature measurement applying thermochromic paintings is based on the specific properties of certain materials, which change color as temperature increases due to changes in their crystal and molecular structure. For example, Arulprakasajothi and Rupesh (2020) applied thermochromic paint, MC 350–8, which is composed of acrylic-silicon resin and pigments. The paint started to change the color at temperatures above 350°C and showed eight different colors as temperaure increased up to 1,270°C.

1.2.2.2 Noninvasive Techniques of Temperature Measurements in Flame

The drawbacks of invasive temperature measurement techniques such as interference of measuring volume promoted the introduction and development of noninvasive techniques of temperature measurement in flame. Within the noninvasive techniques, the significal role belongs to optical tempearture measurement techniques such as infrared pyrometry, Raman scattering, CARS (Coherent Anti-Stokes–Raman) scattering, laser-induced fluorescence (LIF), and laser-induced phosphorescence (LIP) (Leipertz, Pfadler, and Schießl 2010).

The infrared pyrometer measurement utilizes the following physical principle: the flame temperature is determined via measurement of flame radiation intensity at the specified wavelengths in infrared region (about 4.3 μm), where combustion products such as CO_2 exhibit the highest emissivity and strongly affect the thermal radiation intensity (Char and Yeh 1996). The infrared pyrometry has the following advantages: noninvasive measurement without disturbing flow field, flame temperature measurement in wide range from 200°C to 1,650°C ("Optris GmbH"), short response time, and low costs in relation to LIF or CARS techniques. The main disadvantage is the need for precise selection of detector wavelength suitable for specific application since the medium between measuring point and detector may absorb and scatter infrared intensity, which may decrease the accuracy of measurement (Char and Yeh 1996).

In Raman scattering and CARS techniques, the photons are sent by the laser beam to the measured volume, where different molecules exhibit the change in energy level forcing the emitted photon for frequency shift. In combustion systems, Raman scattering is applied for determination of species concentration since almost all species in combustion have Raman-active molecular energy transition (Leipertz, Pfadler, and Schießl 2010). Nevertheless, the temperature measurements in flame could also be performed by Raman scattering technique since the Raman spectrum of the molecules is also temperature-dependent (Lapp and Penney 1979). CARS belongs to nonlinear Raman technique, where three laser beams are irradiated generating four emitted laser beams, which increase sensitivity of high temperature measurements (Marrocco 2010).

LIF is a resonant technique generating stronger signal compared with nonreasonant Raman scattering. LIF technique is based on the absorption–emission relationship between laser photons and molecules in the measured volume. The LIF technique is very selective in measuring the concentration of particular species due to possibility to choose appropriate excitation wavelength (Leipertz, Pfadler, and Schießl 2010). LIF technique may also be applied for the determination of flame medium temperature maps, since excitation spectra give information about population distribution of the molecules (Lawitzki et al. 1990).

1.2.3 FLAME STABILIZATION

The improvement and control of flame stability during FSD process are a crucial issue, since flame instabilities may increase the risk of mechanical damage of dryer elements working in the combustion zone (Jochei et al. 2019), incomplete combustion due to limited mixing and burning rate of fuel and oxygen, which results in increased fuel consumption and generation of pollutants, problems with control of final product properties.

1.2.3.1 Flame Stabilization Problems

In the combustion process, the following flame stabilization problems are distinguished:

- *Oscillation.* Flame instability problems caused by flow oscillation are induced by the natural convection occurring due to density difference between hot gases in the flame zone and cold gases surrounding the flame (Buckmaster and Peters 1988). This natural convection results in the formation of recirculating flow, which interacts with the flame front in axial and radial direction (Jocher et al. 2019). Another cause of combustion instabilities results from heat release oscillations during combustion process and acoustic pressure fluctuations, which is also known as Rayleigh criterion for combustion instabilities (Huang and Yang 2009). The Rayleigh criterion claims that the amplitude of pressure oscillation increases when heat release occurs in the same moment as pressure fluctuation reaches its peak. To mitigate the pressure oscillations, the heat release should take place when pressure oscillation is in minimal point (Rayleigh 1945). In liquid fuel combustion, the fluctuations of heat release, i.e., periodic character of exothermic combustion reaction, are caused by atomization and droplet vaporization, which is slower than chemical reaction rate; instabilities in fuel mass flow rate; flame surface modifications during the process; variations in the local fuel–air ratio; and hydrodynamic instabilities (Huang and Yang 2009).
- *Lift-off height.* The lifted flames are frequently used in industrial processes, where flame should be separated from the fuel atomizing device, i.e., nozzle, to minimize the risk of equipment damage (Min and Baillot 2012). Flame lift-off height is a distance between the burner or atomizing device and the flame base; if the lift-off height reaches critical values, the blowout of flame occurs and flame extinguishes. The lifted flame may be formed, if in the vicinity of nozzle, the access to oxygen is limited, then longer time is required for the formation of stoichiometric mixture of fuel and oxygen; therefore, the stoichiometric mixture and flame are generated at higher distance from the nozzle (Kashir, Tabejamaat, and Baig Mohammadi 2012).
- *Blowout.* The flame blowout occurs if the lift-off limit is exceeded, if the average flow velocity of fuel/oxygen stream is higher than the flame burning velocity (Min and Baillot 2012), or if fuel concentration is not sufficient.
- *Flameout.* The flameout occurs when flame is extinguished in the combustion chamber.

1.2.3.2 Passive and Active Flame Stabilization Techniques

To overcome the flame instability problems, two type of flame stabilization techniques may be applied: active and passive flame stabilization methods, both types of methods are aimed to interrupt the coupling between acoustic and heat release oscillations (Huang and Yang 2009). Passive methods are inexpensive and technically simpler than active methods of flame instability suppression (García-Armingol et al. 2016). Passive flame stabilization methods are targeted to decrease unsteady fluctuations and include the following solutions (Huang and Yang 2009):

- Application of acoustic dumpers such as resonators and acoustic slots to dissipate the energy of oscillations (Richards, Straub, and Robey 2003);
- Application of fuel staging and pilot flames, which implicates separating the main fuel stream into several nozzles to generate two regions: the main combustion zone with excess of oxygen (low equivalence ratio) and region with high fuel concentration able to sustain stable combustion in the main region and avoid flame blowout (García-Armingol et al. 2016). The most typical implementation of fuel staging is application of pilot flames surrounding the main flame region (Riazi et al. 2010). For example, in majority of combustion chambers applied for flame spray pyrolysis (FSP), the main flame is stabilized by pilot flames generated by gaseous fuel supplied into the concentric region surrounding precursor atomization nozzle, (Mueller, Mädler, and Pratsinis 2003). The disadvantages of pilot flame applications are possible increase of NO_x and CO pollution due to the presence of fuel-rich zones; in this case, the proper location and design of pilot flame injectors are required (García-Armingol et al. 2016).
- Improvement of fuel/air mixing is aimed to eliminate the variations in fuel–air ratio, which cause flame instability problems. This flame stabilization technique includes such modifications of combustion system as different configuration of number and location of fuel injection points as well as air inlets to ensure enhanced mixing of fuel and air (Scarinci, Freeman, and Day 2004).
- Modification of combustion chamber and fuel nozzle geometry. The construction of drying chamber could also be changed to ensure stable combustion, for instance, application of air inlets tangential to dryer walls to generate swirl flow of the air for flame stabilization (Yi and Axelbaum 2014). Within the combustion chamber, the bluff body may be installed to promote recirculation of hot combustion products around fuel nozzle (Dayton and Foust 2020). Elliptic-shape fuel nozzle to generate swirl-stabilized fuel–air flow was applied to reduce flame instabilities and decrease generation of NO_x and CO pollutions (Paschereit and Gutmark 2004).

Active methods of flame stabilization involve application of external energy source in order to eliminate combustion instabilities (Huang and Yang 2009). For example, Knoop, Culick, and Zukoski (1997) applied pulsed injection of secondary flame into the recirculation zone of the combustion chamber to stabilize combustion oscillations. Jocher et al. (2019) proposed application of magnetic field for flame stabilization, which decrease the density gradient between hot combustion gases and cold inlet gases reducing the flame oscillations caused by natural convection.

The drawback of passive stabilization methods is limitation to a narrow range of operational process parameters and modifications of combustion chamber geometry, which requires extensive and costly tests of different combustion chamber designs (Huang and Yang 2009). The combustion instabilities are often caused by several phenomena; therefore, the combination of a few flame stabilization techniques both passive and active is required to ensure stable combustion process.

1.3   FLAME SPRAY DRYER

All the experiments on FSD currently available in the literature were carried out in pilot plant installation designed and built at the laboratory of the Faculty of Process and Environmental Engineering, Lodz University of Technology (Figure 1.3), (Piatkowski, Taradaichenko, and Zbicinski 2014; Sobulska, Piatkowski, and Zbicinski 2017). Drying tower (9 m high and 0.5 m in diameter), the upper part of which is shown in the Figure 1.4, is equipped with feeding system, air supplying system, product collecting, and dedusting system. The installation is capable of operating as a conventional spray dryer or flame spray dryer with the drying air heating system shut off. Additionally, the drying tower is armed with ignition system, camera for the flame detection, and safety membranes, which should allow for release of built-up pressure in the case of explosion.

Before drying, a solution or suspension is mixed with a liquid flammable component in a 40 L thermostatic tank (Figure 1.5). The tank is heated by belt-type electric heater installed at the outer wall. The feed is continuously stirred by the blade-type agitator with rotational speed ca. 80 rpm. Once the substrates are dissolved completely, the rotational speed can be reduced to about 30 rpm to limit aeration of the solution, reduce heat losses to the environment, and minimize evaporation of the fuel from the solution.

Solution prepared in the tank is delivered to the nozzle by the high-pressure piston pump. The spraying nozzle in FSD must meet specific requirements to protect atomized material from high-temperature environment. The nozzle head installed in the axis of spray tower is equipped with cooling jacket to control the temperature of the sprayed material (Figure 1.6).

Piatkowski, Taradaichenko, and Zbicinski (2014) tested different spray ignition systems, i.e., ceramic heater (Figure 1.7a), pilot flame (Figure 1.7b), and electrode ignition system (Figure 1.8). After extensive tests, electrode ignition system for the flame ignition was applied due to more stable spray combustion.

The air needed for combustion reaction is supplied downward, co-currently with material flow to the dryer at the ambient temperature through a system of meshes to make the velocity profile uniform. Just after the dryer outlet, particles of the dried material are separated in cyclones and then directed to the bag filter. Finally, after passing through a bag filter, clean drying gas is cooled and removed to the atmosphere. Air flow rate in the dryer was determined experimentally to deliver the amount of oxygen required to sustain the combustion of fuel–lean mixture and to decrease the flame temperature. During tests of FSD process, the equivalence ratio (stoichiometric air/fuel ratio to the current air/fuel ratio) was in the range from ca. 0.03 to 0.3, which corresponds to an excess of oxygen in the drying tower – current

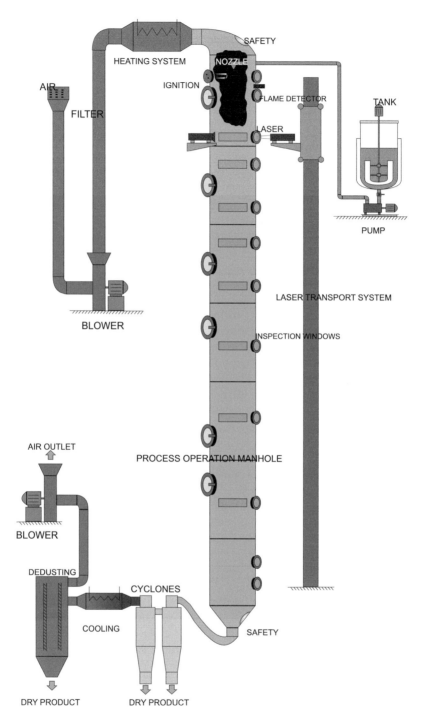

FIGURE 1.3 Experimental installation for FSD process tests – scheme. (Sobulska, Piatkowski, and Zbicinski 2017, Permission Granted).

FIGURE 1.4   Photograph of the upper part of the flame spray drying tower.

FIGURE 1.5   Photograph of thermostatic tank used for feed preparation.

FIGURE 1.6   Picture of pressure nozzle.

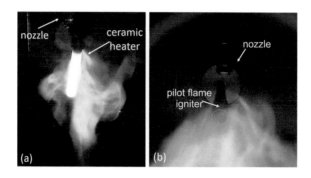

FIGURE 1.7   Picture of ceramic heater (a) and pilot flame (b) ignition elements.

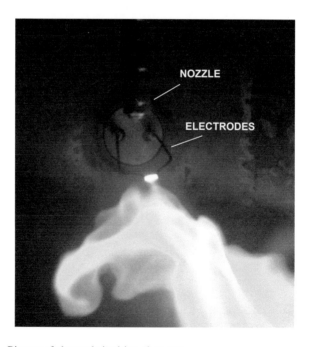

FIGURE 1.8   Picture of electrode ignition elements

air/fuel ratio ranged from ca. 30 to 300, whereas stoichiometric air/fuel ratio of ethanol is 9.

Figure 1.9 shows the flame in the drying tower during FSD of maltodextrin solution with the addition of ethanol. The shape and size of the flame are similar to the shape and size of the atomization envelope observed in a conventional spray drying process under comparable conditions, i.e., similar feed and air flow rates.

To control the FSD process, drying tower has been equipped with control and measuring sensors connected to the measuring control system operating in an autonomic mode (Sobulska 2019).

In front of the inspection windows, the digital camera has been installed to ensure the detection of the flame and control of the combustion process. In the case when, despite attempts of firing the jet, the flame is not detected, the pump supplying the feed to the nozzle is shut down to avoid an uncontrolled spread of fire and explosion in the dryer. Flame control and safety membranes made from thin aluminum sheets installed on the top and bottom of the dryer allowed safe running of FSD process.

The measurements of particle size distribution (PSD) and velocity distributions were carried out using laser measurement techniques, i.e., particle dynamics analysis (PDA) and laser Doppler anemometry (LDA) systems, which allows for nonintrusive in situ measurements of velocity (one to three components) and size of the spherical particles.

Figure 1.10 shows PDA laser measurement in flame, where crossing of the laser beams indicates measuring point.

To determine the mechanism of FSD process, Sobulska, Piatkowski, and Zbicinski (2017) measured the local particle size and velocity distributions at the distances of

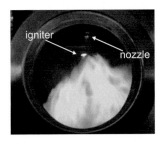

FIGURE 1.9 View from the side inspection window of the tower. (Piatkowski, Taradaichenko, and Zbicinski 2015 Permission Granted).

FIGURE 1.10 Laser measurement in the flame (transmitter and crossing of laser beams).

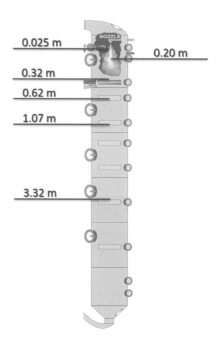

FIGURE 1.11   Levels for laser measurements.

0.14, 0.20, 0.32, 0.62, 1.07, and 3.32 m from the nozzle with ignition of the sprayed stream (Figure 1.11). Due to short flame length, the measuring points are situated in the upper part of the drying tower to ensure accurate analysis of particles sizes and velocities in the combustion zone. Lower measuring points are used to evaluate change of particle size and particle velocity after leaving the combustion zone.

Accuracy of droplets and particles size measurement using laser techniques depends on the number of droplets that are collected at the measuring point. In the FSD experiments in each measuring point, the PDA system collected from 1,000 to 15,000 droplets/particles with diameters from 1.7 to 1,000 μm, so after Lefebvre (1988), an average error of the laser measurements can be estimated as 5%–10% (Lefebvre 1988).

To develop safe and energy-efficient FSD process, Piatkowski, Taradaichenko, and Zbicinski (2014) analyzed the flame stability limits.

The stability limits of the flame in FSD process may be affected by the following parameters: the fuel concentration in the feed, feed rate, atomization parameters, air/fuel ratio, and air velocity in the drying tower. The concentration of the fuel in the feed should be low to reduce energy and fuel consumption; however, the amount of fuel in the feed should be high enough to generate stable flame and produce completely dried final product. Air flow rate in the drying tower should be fixed on the level to supply a sufficient amount of oxygen into the combustion/atomization zone without blowing off the flame.

According to Yi and Axelbaum (2014), the flame stability of the high-water-content alcohol fuels depends on the preferential vaporization of fuel from the droplets: for example, due to high volatility and high activity coefficient of alcohols in aqueous solution, the stable combustion of alcohol fuels may be achieved even for fuels with high water content. The authors determined the minimal fuel concentration needed to achieve stable combustion of water/ethanol spray in air of 25 wt.% and in oxygen 10 wt.% (Yi and Axelbaum 2013). For t-butanol and 1-propanol, which have higher activity coefficient compared with ethanol, the minimal concentration in water/fuel mixture needed to obtain stable flame in air was about 17.2 wt.% for 1-propanol (Yi and Axelbaum 2013) and ca. 12.5 wt.% for t-butanol (Yi and Axelbaum 2014). The significant decrease of butanol concentration needed for stable combustion was explained by higher evaporation rate of t-butanol, theoretical calculations showed that the time needed for evaporation of 99% of fuel from the droplet was over 70% shorter for t-butanol compared with ethanol (Yi and Axelbaum 2014). Figure 1.12 shows the blow-off limits (the lowest concentration of oxygen where a flame is stable under a given flow rate of N_2/O_2 mixture) for ethanol/water (10 and 25 wt.%), 1-propanol/water (17.2 wt.%), and t-butanol/water spray combustion (12.5 wt.%). The figure shows that stable combustion of fuel with even 90% of water and only 10% of ethanol may be obtained by increase of oxygen concentration over 60%.

The aim of the research of Piatkowski, Taradaichenko, and Zbicinski (2014) on flame stability limits was to reduce ethanol concentrations in the raw material to increase energy efficiency of FSD. Figure 1.13 shows the flame stability map in FSD process determined as a minimal ethanol concentrations in the raw material needed to obtain stable flame for nozzle with orifice diameter 0.51 mm, solids content in the feed 12 wt.%, and wide range of air flow rates from 50 to 500 Nm^3/h and feed rates from 3 to 10 kg/h.

Combustion of the sprayed material is stable when, after ignition of the spray, the flame could sustain without igniter assistance for the period of minimum 5 min. For fuel concentration of 35 wt.%, the stable flame was sustained within the limited range

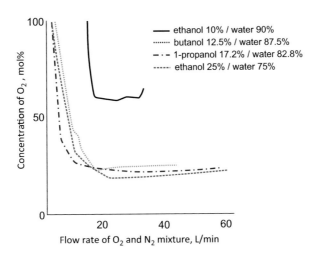

FIGURE 1.12 Blow-off limits for ethanol/water, 1-propanol/water, and t-butanol/water spray combustion. (Based on Yi and Axelbaum 2014 and Yi and Axelbaum 2013.)

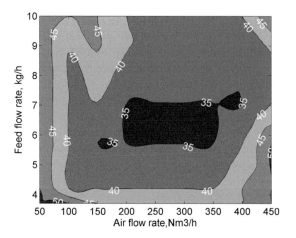

FIGURE 1.13   Flame stability map of FSD (solid content of 12 wt.%, experimental data) – the minimal concentration of fuel (ethanol) in the feed needed to sustain stable combustion at the different feed rates and air flow rates in the drying tower.

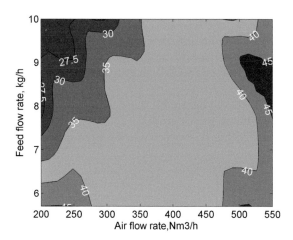

FIGURE 1.14   Flame stability map of FSD (solid content of 32.5 wt.%, experimental data) – the minimal concentration of fuel (ethanol) in the feed needed to sustain stable combustion at the different feed rates and air flow rates in the drying tower.

of process parameters: feed flow rate from 5.7 to 7.1 kg/h and air flow rate from ca. 150 to 400 Nm^3/h, Figure 1.13.

Increase of solid content in the feed to 32.5 wt.% allowed for reduction of the fuel concentration in the initial solution: the concentration of fuel (ethanol) of 35 wt.% could be applied in the wide range of process parameters: air flow rate from 250 to 550 Nm^3/h and feed flow rate from 5.7 to 10 kg/h (nozzle with orifice diameter 0.41 mm) – Figure 1.14.

The authors showed also the FSD regime where fuel concentration in the feed could be decreased even to 27.5 wt.%: for feed flow rate of 9–10 kg/h and for air flow rate from 200 to 300 Nm^3/h. Decrease of fuel concentration in the feed from 35 to

27.5 wt.% allows for reduction of specific energy consumption for about 28% (from 28.53 to 20.43 MJ/kg H_2O), which significantly enhances of FSD process efficiency (Lodz University of Technology, 2015, unpublished data).

1.4 FLAME SPRAY DRYING EXPERIMENTS

Sobulska (2019) described the mechanism of FSD based on extensive drying tests of thermally resistant and thermally sensitive materials, i.e., ceramic material, maltodextrin, and coffee. To establish the mechanism of FSD, the authors analyzed the effect of different FSD process parameters such as feed compositions (content of flammable component and solid), feed flow rate, atomization pressure, nozzle type, feed temperature, and air flow rates on the properties of dispersed phase (particle size and velocity distribution and degree of thermal degradation).

Maltodextrin solutions with different solid contents in the fuel-free solution were from 30 to 50 wt.%, whereas coffee and ceramic material with solid content in the fuel-free solution of 50 wt.% were dried (Table 1.2).

The authors applied two flammable components for FSD tests: ethanol and sunflower oil. Ethanol is a biofuel commonly used in motor industry, mainly as an additive for gasoline. Vegetable oils such as sunflower oils are an alternative type of biofuels, which can be used directly in the diesel engines or can be processed into biodiesel. Both fuels selected for FSD tests: ethanol and sunflower oil, are nontoxic, chemically neutral to the dried materials, with high boiling point and high heating values: sunflower oil – 38 MJ/kg (Ryan, Dodge, and Callahan 1984) and ethanol – 26.85 MJ/kg (Green and Perry 2008). During the FSD tests, the concentration of ethanol in the feed varied from 25 to 45 wt.%, and the concentration of sunflower oil in the feed was constant and equal to 50 wt.% (Table 1.2).

A wide range of atomization pressures from 0.3 to 6.2 MPa were applied (Table 1.3) to analyze the effect of atomization pressure on the temperature distributions, product properties, and particles hydrodynamics in the drying tower.

Due to transient character of the combustion process, the oscillations of air pressure in the tower in the range from ca. 260 to 450 Pa were observed. Figure 1.15

TABLE 1.2
Feed Compositions Applied During FSD Tests

Product	Solids Content in the Fuel-Free Solution (wt.%)	Fuel Concentration in the Feed (wt.%)	Fuel Type
Maltodextrin	30	35, 40, 45	Ethanol
Maltodextrin	40	35, 40, 45	Ethanol
Maltodextrin	50	27.5, 35	Ethanol
Maltodextrin	30, 40, 50	0	Standard spray drying
Ceramic material	50	25, 50	Ethanol
Ceramic material	50	50	Sunflower oil
Ceramic material	50	0	Standard spray drying
Coffee	50	45	Ethanol

TABLE 1.3

Range of Process Parameters Applied During FSD Tests

Solution	Atomization Pressure (MPa)	Feed Rate (kg/h)	Feed Temperature (°C)	Air Flow Rate (N m³/h)	Nozzle Type[a]
Maltodextrin (30 wt.%)	6.2, 0.9	5.5, 10	70, 30	250	Fine 0.6 Cone jet 0.6
Maltodextrin (40 wt.%)	5.3, 5.0, 1.6, 0.7, 0.3	5.5, 7, 10	70	340, 400, 480	Fine 0.4, 0.6, 1.0 Cone jet 0.6
Maltodextrin (50 wt.%)	4.1, 3.9, 2.5	10	70	230, 360	Fine 0.6
Ceramic material	0.9, 0.8	10	70	200, 250	Fine 1.0
Coffee	0.7, 1.7	10	70	382, 395	Fine 0.6 Cone jet 0.6

[a] Spraying Systems, USA.

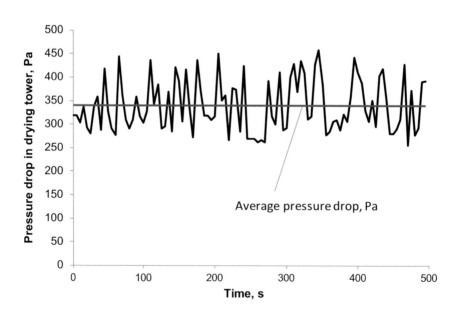

FIGURE 1.15 Pressure in the tower recorded during 500 s of measurement (under pressure).

illustrates oscillations of air pressure in the tower recorded during FSD process (500 s of measurements).

1.5 DROPLETS AND PARTICLES FLUID DYNAMICS

The mechanism of simultaneous combustion and drying of liquid droplets during FSD process, the effects of feed composition, feed flow rate, atomization pressure, nozzle type, feed temperature, and air flow rates on the particle size and velocity distribution have been described in Sobulska, Piatkowski, and Zbicinski (2017) and Sobulska, Zbicinski, and Piatkowski (2020).

The most important findings are summarized in sections 1.5–1.6.

1.5.1 INITIAL PARTICLE SIZE AND VELOCITY DISTRIBUTION

Determination of initial particle size (PSD) and particle velocity distribution in spray envelope is the base to describe the mechanism of FSD.

- Effect of atomization pressure

  Figure 1.16 shows initial PSD for different atomization pressure 5.3, 1.6, and 0.7 MPa. Decrease of atomization pressure produces higher arithmetic mean diameters of particles (AMD) – 23, 30, and 34 μm and higher Sauter Mean Diameters (SMD) – 59, 97, and 105 μm and smaller spraying angle – 80°, 75°, and 62° resp. (Table 1.4).

  Figures 1.17 and 1.18 illustrate the AMD determined at different positions along the drying tower radius for FSD tests with atomization pressure 5.3 MPa (Fine spray) and 1.6 MPa (Coarse spray). For coarse spray, initial

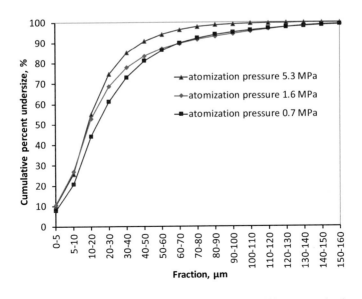

FIGURE 1.16   Initial cumulative droplet size distribution for different atomization pressure.

TABLE 1.4

Spray Parameters for Different Atomization Pressure

Atomization Pressure (MPa)	Initial Droplets AMD in the Spray (µm)	Initial Droplets SMD in the Spray (µm)	Spraying Angle (°)
0.7	34	105	62
1.6	30	97	75
5.3	23	59	80

Source: Sobulska, Zbicinski, and Piatkowski (2020). Permission Granted.

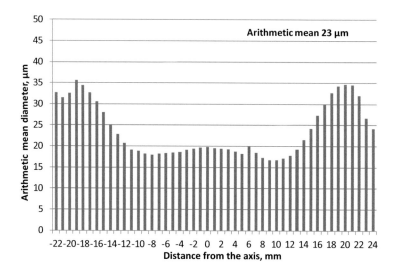

FIGURE 1.17 AMD determined at different positions along the drying tower radius for fine spray – atomization pressure 5.3 MPa (Sobulska, Piatkowski, and Zbicinski 2017. Permission Granted).

droplet AMD ranges from 15 µm at the axis of the spray to 51 µm at the spray edge. In fine sprays, smaller droplets with diameters from 17 µm at the spray axis to 35 µm at the edge of the spray have been found. Initial PSD presented in Figures 1.17 and 1.18 is typical for hollow cone pressure nozzle atomization.

Figures 1.19 and 1.20 show the initial average droplet velocities of particle fractions along the drying tower radius for fine (atomization pressure 5.3 MPa) and coarse spray (atomization pressure 1.6 MPa). For high atomization pressure (Figure 1.19), similar velocities (c.a. 18 m/s) of all the fractions in the core of the spray were observed. The pattern changes at the spray edge; the larger the fraction, the higher is the velocity. For lower

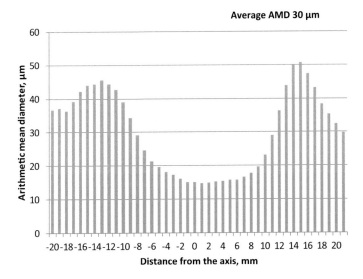

FIGURE 1.18 AMD determined at different positions along the drying tower radius for coarse spray – atomization pressure 1.6 MPa.

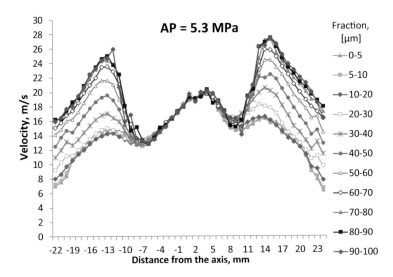

FIGURE 1.19 Initial average droplet velocities of particular fractions at different positions along the drying tower radius for fine spray – atomization pressure 5.3 MPa. (Sobulska, Piatkowski, and Zbicinski 2017 Permission Granted.)

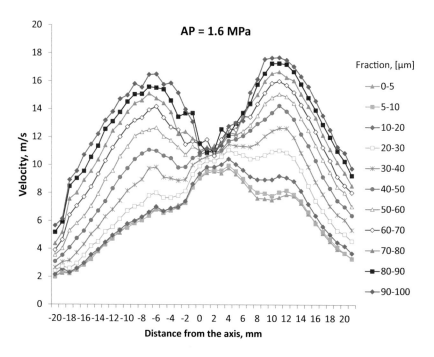

FIGURE 1.20 Initial average droplet velocities of particular fractions at different positions along the drying tower radius for coarse spray – atomization pressure 1.6 MPa.

initial atomization pressure (1.6 MPa, coarse atomization, Figure 1.20), pattern of particle velocities in the spray is similar to that in the fine atomization; however, average droplet velocities are smaller (9–11 m/s) due to lower kinetic energy of the spray. Distribution of the average droplet velocities presented in Figures 1.19 and 1.20 is typical for the fine pressure nozzle atomization.

- Effect of feed rate

  The initial PSD for different feed rates is shown in Figure 1.21. In general, for pressure nozzles, decrease of feed rate results in decrease of atomization pressure, which produces larger initial droplet diameters, smaller spraying angles, and lower number of droplets in the spray.

- Effect of solid content

  Due to increase of solid content and, as a result, higher viscosity of the solution, large initial droplet diameters are produced, i.e., higher AMD, SMD, and smaller spraying angle (Figure 1.22, Table 1.5).

- Effect of fuel concentration

  The effect of fuel concentration in the feed on the initial PSD is shown in Figure 1.23. Initial cumulative PSD is shown for tests with fuel content of

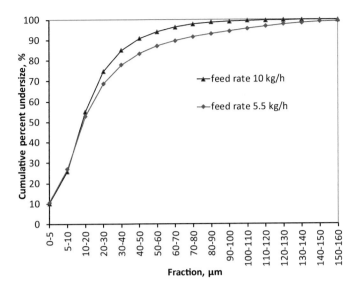

FIGURE 1.21 Initial cumulative droplet size distribution for different feed rate: 5.5 and 10 kg/h.

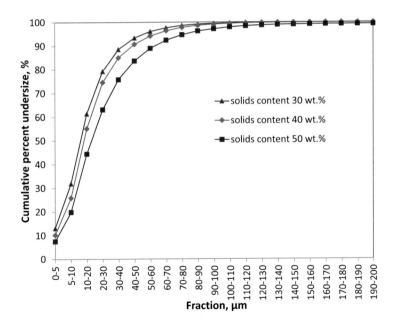

FIGURE 1.22 Initial cumulative droplet size distribution for different solid content: 30, 40, 50 wt.%.

TABLE 1.5

Initial Droplet AMD and Spraying Angle for Different Solid Content in the Feed: 30, 40, 50 wt.%

Solid Content in the Feed (wt.%)	Initial Droplets AMD (µm)	Initial Droplets SMD (µm)	Spraying Angle (°)
30	21	52	82
40	23	59	80
50	33	71	76

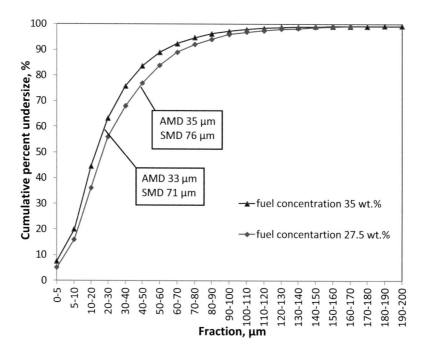

FIGURE 1.23 Initial cumulative PSD for different fuel concentration in the feed 27.5 and 35 wt.%.

27.5 and 35 wt.%. Decrease of the fuel concentration in the feed increases the viscosity of the solution, which slightly increases initial droplet diameters.

- Effect of feed temperature on initial PSD

    The reduction of feed temperature results in increase of the initial droplets AMD due to lower viscosity of sprayed solution (Figure 1.24).

    Comprehensive information about initial atomization parameters obtained in the experiments constitutes a database for the analysis of the mechanism of FSD.

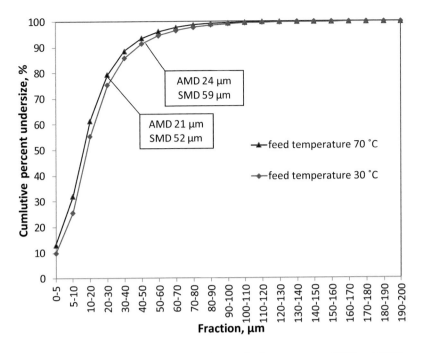

FIGURE 1.24   Initial cumulative PSD for different feed temperatures 30°C and 70°C.

1.5.2   Particles and Droplets Flow Pattern

Comparison of local PSDs at different distances from the nozzle in FSD and SSD process for similar feed rate, nozzle type, atomization pressure, air flow rate in the tower, and similar outlet air temperatures is presented in the Figures 1.25 and 1.26 resp.

In SSD process, substantially different PSD in radial locations in comparison to FSD is observed. In FSD process (Figure 1.25), uniform PSD already at the level 0.62 m below the flame is formed, which proves that flow turbulence generated by the combustion process caused intensive particles mixing. Similar conclusions were reported in Gounder, Kourmatzis, and Masri (2012), in the study of droplet flow dynamics and turbulence intensity in ethanol and acetone spray flames; the authors reported that the highest turbulence intensity was found at the outer regions of the flame.

Figures 1.27a and b show the average particle velocities along the tower radius for different distances from the nozzle for SSD and FSD. Particle velocity distributions along the tower radius are not uniform at all the levels below the nozzle for both SSD and FSD. At the distance of 3.32 m from the nozzle, profile of the velocity in the cross section of the drier flattens out and particle velocities become close to air velocity. For FSD process (Figure 1.27b), negative particle velocities at the edges of the flame at distance of 0.2 m from the nozzle are detected, which indicates the development of recirculation zones and promotion of particle agglomeration.

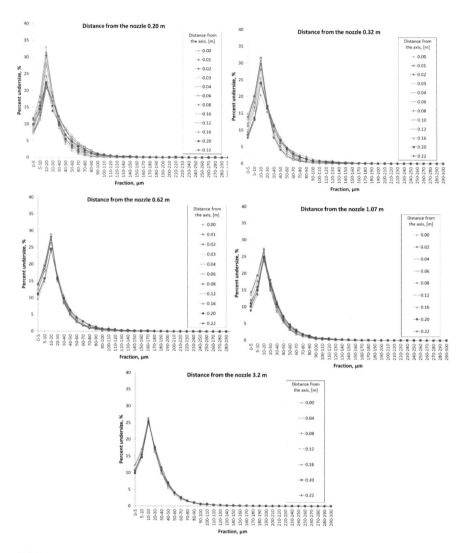

FIGURE 1.25 Local PSD for different distances from the nozzle in flame spray drying process.

For standard concurrent spray drying process, no negative particle velocities are observed (Figure 1.27a) due to parallel flow of the continuous and dispersed phases (Zbiciński and Piątkowski 2004).

- Effect of atomization pressure

    Figures 1.28 and 1.29 display changes of AMD and SMD along the drying tower height for different atomization pressures in FSD process.

    In each FSD test particle sizes increase with distance from the nozzle due to particle agglomeration and puffing.

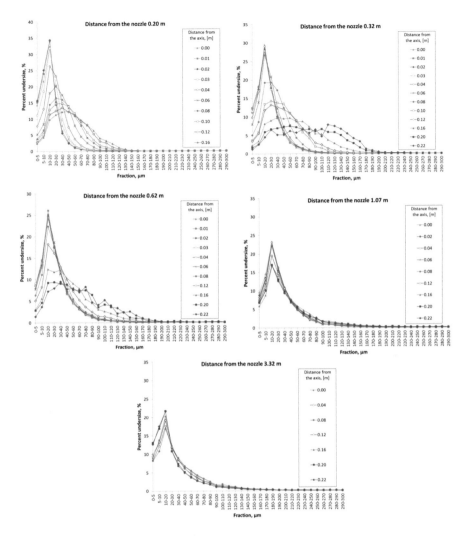

FIGURE 1.26   Local PSD for different distances from the nozzle in standard spray drying process.

For the fine spray (5.3 MPa), the flame length is ca. 0.20 m, whereas for coarse spray (0.7 MPa), flame length is longer, 0.32–0.62 m, which increases particle residence time in the combustion zone and might result in the formation of puffed or inflated particles. Sano and Keey (1982) studied puffing of skim-milk particles during spray drying and reported that if drying air temperature exceeded the boiling point of the solvent, inflation of droplets occurred (Sano and Keey 1982). The authors showed that particle inflation resulted in the acceleration of drying rate due to increased drying surface and reduced crust thickness. Experimental study of single droplet drying kinetics of aqueous lactose solution of Tran et al. (2017) showed that

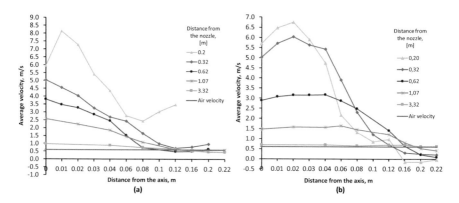

FIGURE 1.27 Average particle velocities at different distances from the nozzle for (a) standard spray drying and (b) flame spray drying.

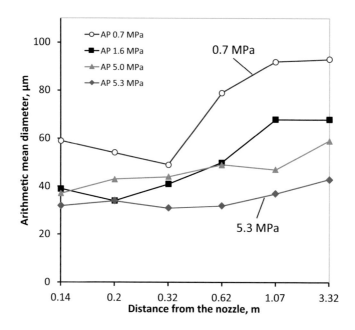

FIGURE 1.28 Changes of AMD along the drying tower height for different atomization pressure.

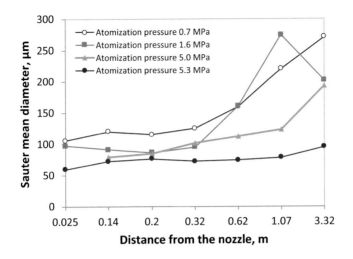

FIGURE 1.29 Changes of SMD along the drying tower height for different atomization pressure.

at temperatures above 100°C, internal boiling of droplet took place with bubble formation resulting in droplet inflation/deflation. Longer flame and higher droplet residence time in the flame for spray with coarse droplets were also reported in the work of Heine and Pratsinis (2005), who analyzed the FSP process.

Figure 1.30 shows the average particle velocities along the tower radius and at different distances from the nozzle for different atomization pressures. For atomization pressure of 6.2 MPa (Figure 1.30a), particle velocity distribution along the tower radius is not uniform from 0.14 to 0.62 m below the nozzle. At the distance from the nozzle of 3.32 m, the velocity profiles in the cross section of the drier are flat and particle velocities become close to drying air velocity of 0.4 m/s. For lower atomization pressure, 0.9 MPa, similar patterns of the average particle velocities are observed (Figure 1.30b). Analysis of the results shows that the negative values of the particle velocities occur at the edges of the flame developing recirculation zones, which enhance particle agglomeration during FSD process.

- Effect of feed rate

Figure 1.31 shows the AMD distributions along the radius at different vertical positions of the tower for different feed rates of 5.5 and 10 kg/h. For lower feed rate (Figure 1.31a), the AMD at the axis of the tower increases up to 70 μm, whereas at the edges of the flame reaches even 70–140 μm, which is significantly higher than the initial values (15–45 μm), due to particle agglomeration process.

The particle AMD in the drying tower obtained for feed rate 10 kg/h (Figure 1.31b) is lower: at the level 0.2 m from the nozzle, AMD varied from 20 μm at the axis of the tower to 80 μm at the edge of the flame.

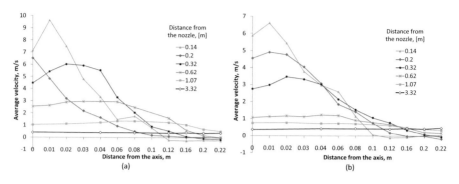

FIGURE 1.30 Average particle velocity for different positions along radius and height of the drying tower for solid content of 30 wt.% and atomization pressure: (a) 6.2 MPa; (b) 0.9 MPa (Sobulska, Piatkowski, and Zbicinski 2017 Permission Granted).

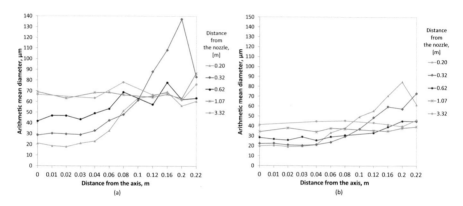

FIGURE 1.31 Effect of feed rate on the AMD in the drying tower for feed rate: (a) 5.5 kg/h, (b) 10 kg/h.

Increase of particle AMD for lower feed rate is caused by reduction of atomization pressure, which produces longer flame, higher flame temperatures, and longer residence time of particles in the flame, which results in particle puffing.

- Effect of solid content in the feed

  Sobulska et al. (2020) determined the particle flow pattern during the FSD at different locations of the drying tower for different solid content in the feed (Sobulska, Piatkowski, and Zbicinski 2017). Figure 1.32 shows the increase of particle AMD in drying tower for different solid contents from 37 to 64 μm along the distance from the nozzle.

  Increase of particle diameters for higher solid content may be explained by particle puffing due to longer residence time in high-temperature combustion zone. Increase of solid content in the feed generates higher flame temperature and longer flame (from 0.2 to 0.62 m) (Section 1.6). Longer flame length for high solid content results from higher viscosity of the

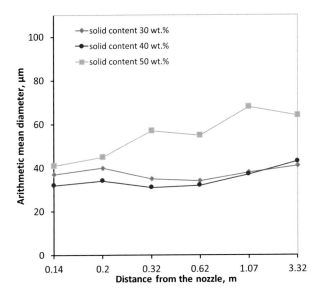

FIGURE 1.32 Changes of AMD along the drying tower height for different solid content: 30, 40, and 50 wt.%.

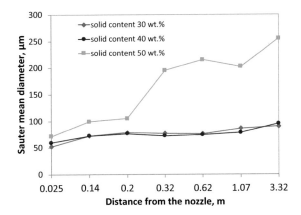

FIGURE 1.33 Changes of SMD along the drying tower height for different solid content: 30, 40 and 50 wt.% (Sobulska, Zbicinski, and Piatkowski 2020 Permission Granted).

solution, which changes the atomization parameters toward generation of coarse spray (larger initial droplet size, small spraying angle, bigger spacing between droplets in the spray). In the coarse spray, enhanced entrance of the drying air into the spray core promotes intensive combustion. Similar effect was observed for SMD (Figure 1.33).

Figure 1.34 shows PSD and particle velocity distribution for different solid content at different positions along the drying tower cross section at distance 0.20 m from the nozzle, which is within the combustion zone. The

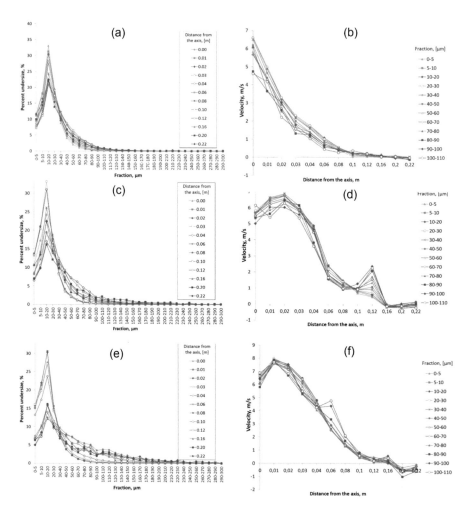

FIGURE 1.34 PSD at the measuring level of 0.2 m from the nozzle for solid content in the fuel-free solution: (a) 30 wt.%, (c) 40 wt.%, (e) 50 wt.%; and velocity distributions for particular fractions at the measuring level of 0.20 m for solutions: (b) 30 wt.%; (d) 40 wt.%; (f) 50 wt.%.

highest number of large particles is observed at the flame edges as a result of negative particle velocities in the recirculation zones. For solid content of 30 wt.%, upward flow of the particles was observed at point 0.22 m from the axis with average particle velocity of −0.014 m/s. When solid content was increased to 50 wt.%, particle recirculation zone covered a larger area: negative velocities of −0.52 and −0.39 m/s were observed at positions 0.20 and 0.22 m.

Figures 1.35a and b show AMD at different positions along the drying tower radius and in different vertical locations for solid content in the feed of 30 and 50 wt.%. Analysis of the figure shows that for higher solid content in the feed, more sharp increase of particle diameters is observed starting

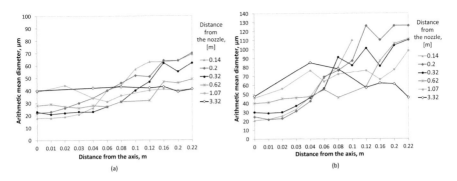

FIGURE 1.35 Effect of solid content in the feed on the AMD in the drying tower: (a) solid content of 30 wt.%. (Sobulska, Zbicinski, and Piatkowski 2020 Permission Granted); (b) solid content of 50 wt.%.

from the level in the vicinity of the nozzle. Increase of particle AMD for high solid content in the feed was caused by particle puffing due to over-heating in long flame and agglomeration caused by particle recirculation.

- Effect of fuel concentration

    Increase of fuel concentration in the feed decreases the AMD of the particles with a distance from the nozzle. The reduction of ethanol content in the feed lowers the viscosity of the solution, resulting in production of higher initial particle diameters.

- Effect of air flow rate

    No effect of air flow rate in the tower on PSD during FSD process was found. The changes of air flow rate have no significant impact on particle AMD and SMD along the drying tower height due to similar initial atomization parameters.

- Effect of feed temperature on AMD

    The lower the feed temperature, the bigger particle diameters are produced, which may be explained by higher viscosity of the feed.

1.6 FLAME TEMPERATURES

Analysis of flame temperature gives valuable information about the intensity of combustion process during FSD as well as provides data on temperature pattern within the dryer. For temperature measurements within the flame, type S thermocouple might be used to obtain high-accuracy measurements in the high-temperature flame. Due to unstable character of the combustion process during FSD, spray temperature must be calculated as an arithmetic average from several minutes of measurements at each measuring point.

Figures 1.36a and b illustrate the amplitude of temperature fluctuations in FSD recorded during 45–60 s measurement. Both figures show temperature fluctuations recorded within the flame at the distance from the dryer axis of 0.04 m at the two different distances from the nozzle: 0.20 m (Figure 1.36a) and 0.32 m (Figure 1.36b). Within the flame, the temperature varies from 100°C to 945°C in the vicinity of

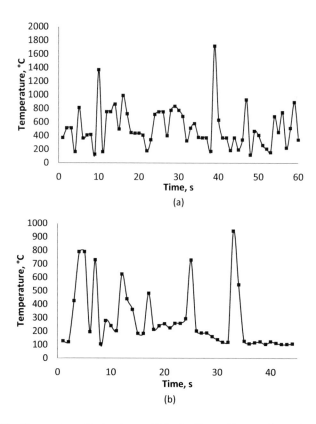

FIGURE 1.36 Temperature fluctuations within the flame, distance from the axis of 0.04 m and distance from the nozzle: (a) 0.20 m and (b) 0.32 m.

the nozzle and from 115°C to about 1,700°C further from the nozzle (0.32 m) in the region where flame is well developed. At the area outside the flame temperature fluctuations decreasing, as an example, Figure 1.37a and b show the change of temperature recorded outside the flame close to the dryer wall (distance of 0.22 m from the dryer axis/0.20 m from the nozzle – Figure 1.37a) and in the region below the flame (distance of 1.07 m from the nozzle/dryer axis – Figure 1.37b). The amplitude of temperature fluctuations in the vicinity of dryer wall is lower (from 41°C to 196°C) compared with dryer axis, Figures 1.36a and 1.37a. At the region below the flame, the temperature oscillates in the smaller range, i.e., from 195°C to 271°C (Figure 1.37b).

The analysis of the FSD process mechanism must deliver information on how FSD process parameters affect the temperature and length of the flame (Sobulska, Zbicinski, and Piatkowski 2020).

- Effect of atomization pressure
  To analyze the effect of atomization pressure on the flame temperature and length, spray temperature should be measured at the different vertical and radial positions in the flame.

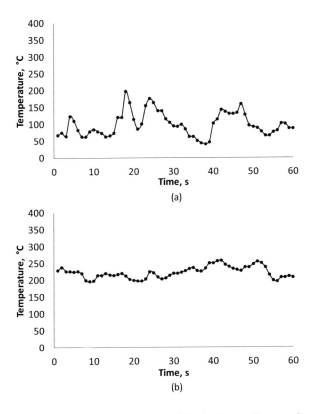

FIGURE 1.37 Temperature fluctuations (a) outside the flame (distance from the axis of 0.22 m and distance from the nozzle of 0.20 m) and (b) below the flame in dryer axis (distance from the nozzle of 1.07 m).

Figures 1.38a–c show spray temperature pattern in the tower for atomization pressure of 5.3, 1.6, and 0.7 MPa, resp. The flame temperatures presented on the figures are average values calculated from temperature data recorded in one measuring point for 2 min.

In all three tests, the concentrations of fuel and solids content in the feed are similar (35 wt.% of ethanol and 40 wt.% of maltodextrin in the fuel-free solution). In FSD for atomization pressure of 5.3 MPa, the range of temperature in the drying tower changes from 52°C to 315°C (Figure 1.38a), flame length does not exceed 0.2 m from the nozzle. For test with atomization pressure of 1.6 MPa (Figure 1.38b), spray stream temperature varies from 73°C to 533°C and flame length from 0.2 to 0.32 m. Reduction of atomization pressure to 0.7 MPa results in an increase of stream temperature: from minimal temperature equal to 106°C and maximal 674°C (Figure 1.38c) and flame length from 0.32 to 0.62 m.

For lower atomization pressure (large droplets in the spray and small spraying angle), longer flame and higher temperatures are observed due to better access of the air into the core of the spray.

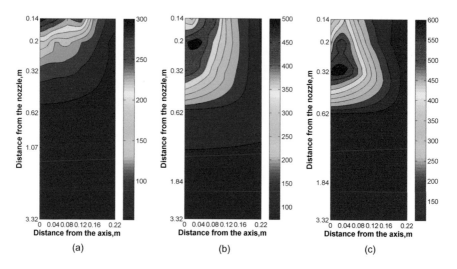

FIGURE 1.38   Experimental spray temperature pattern in the drier for different atomization pressures: (a) 5.3 MPa, (b) 1.6 MPa; (c) 0.7 MPa and solid content 40 wt.%. (Sobulska, Zbicinski, and Piatkowski 2020 Permission Granted.)

Differences in the flame length and combustion intensity for fine and coarse spray can be explained by the group combustion theory introduced by Chiu and Liu (1977). The authors of the theory identified four types of droplet and spray combustion: the isolated-droplet combustion, the internal group combustion, external group combustion, and external sheath combustion. Authors emphasized that position of the flame zone in relation to the individual droplets and spray depends on droplet concentration, droplet diameter, and spacing between droplets.

Four different combustion models were distinguished according to the group combustion number, G:

$$G = 3\left(1 + 0.276\mathrm{Re}^{1/2}\mathrm{Sc}^{1/3}\right)\mathrm{Le}N_d^{2/3}\left(R/l\right) \tag{1.1}$$

For nonmoving droplet (Re = zero), equation (1.1) simplifies to

$$G = 3\mathrm{Le}N_d^{2/3}\left(R/l\right) \tag{1.2}$$

Figure 1.39 presents different regimes of droplet combustion depending on different values of group combustion number, G. Figure 1.39a shows isolated-droplet combustion, which takes place when concentration of the droplets in the spray is low and flame front is formed around each droplet in the spray. In this case, group combustion number G is below 10^{-2}. With an increase of droplet concentration in the spray, the regions with groups of non-vaporized droplets surrounded by internal or external flame start to form. Internal group combustion occurs when the droplets at the edge of the cloud show characteristics of individual droplet combustion with individual

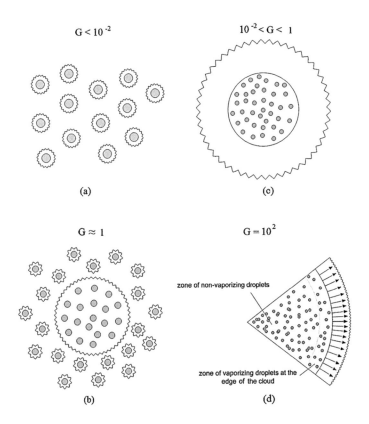

G < 10^{-2} 10^{-2} < G < 1

(a)   (c)

G ≈ 1   G = 10^{2}

zone of non-vaporizing droplets

zone of vaporizing droplets at the
edge of the cloud

(b)   (d)

FIGURE 1.39 Combustion regimes: (a) isolated-droplet combustion, (b) internal group combustion, (c) external group combustion, (d) external sheath combustion. (Based on Sirignano 1999.)

flame front for each droplet. However, in the core of the cloud, the group of droplets share one flame front and, in this case, G is equal or close to unity (Figure 1.39b). If concentration of droplets in the cloud increases and spacing between droplets decreases, the external group combustion is observed (value of G varies between 1 and 10^2), the cloud of droplets is so dense that flame front exists only at the outward regions of the cloud (Figure 1.39c). When G exceeds 10^2, the regime of external sheath combustion is identified, i.e., in the core of the cloud, droplets are non-vaporizing and at the edge of the cloud, droplet evaporation and combustion occur (Figure 1.39d).

Sirignano (1999) mentioned that many industrial burners work in the G regime close to unity characteristic for external group combustion. Imaoka and Sirignano (2005) emphasized that in case of external group combustion, droplets on the edge of the cloud will vaporize more than 5,000 times faster than droplets in the core of the cloud.

In FSD tests, external sheath combustion was observed for fine sprays (high atomization pressure), where dense spray core contained high number of small non-evaporated droplets with small spaces between droplets where access of oxygen was limited, which explains smaller flame length and lower flame temperatures in the drying tower.

When atomization pressure is reduced, large droplet diameters are produced, spaces between droplets increase, which enhanced the entrance of air into the spray core, resulting in increase of the flame temperature and length.

The effect of higher flame temperature for lower atomization pressure and larger droplet diameter in the spray can be even more profound when solution with lower solid content (30 wt.%) is dried. Figure 1.40 shows the temperature pattern determined for FSD with similar fuel concentration (35 wt.%) and solid content, however, for significantly different atomization pressure, i.e. 6.2 MPa (Figure 1.40a) and 0.9 MPa (Figure 1.40b). In this case, decrease of atomization pressure also causes reduction of spray cone and increase of initial droplet diameters: AMD increases from 20 to 26 μm and SMD from 52 to 90 μm. Due to decrease of atomization pressure, the spray atomization type switched from fine atomization to coarse atomization with larger spacing between droplets.

- Effect of feed rate

  The effect of feed rate on temperature pattern in the drying chamber has been already presented in Figure 1.38a and b. Decrease of feed rate from 10 to 5.5 kg/h results in the higher temperatures in the flame zone as lower feed rate lowers atomization pressure, which produces larger droplets and enhances air access into the spray core.

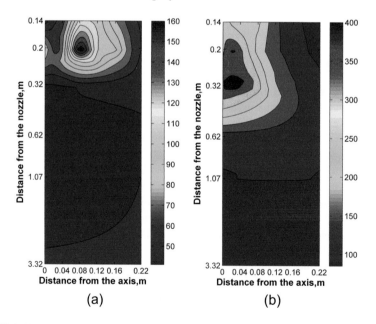

FIGURE 1.40 Experimental spray temperature pattern in the drier for different atomization pressures: (a) 6.2 MPa, (b) 0.9 MPa, fuel concentration (35 wt.%).

- Effect of solid content

To illustrate the effect of solid content in the feed on combustion intensity and flame length, the temperature of the spray is shown at the different positions in the drier cross section and in different distances from the nozzle.

Figure 1.41 presents spray temperature pattern for FSD for different feed solid content. Analysis of Figure 1.41 shows that with increase of solid content in the feed, flame temperature in the drying tower also increases. For feed solid content of 30 wt.%, Figure 1.41a and atomization pressure of 6.2 MPa, high temperatures, up to 180°C, are observed only at the side edges of the spray, whereas in the spray core, temperature decreases to 45°C–74°C due to evaporation from droplets. As solid content in the feed increases to 40 and 50 wt.%, the atomization pressure decreases to 5.3 and 4.1 MPa and the high spray temperatures can be observed at the spray edges and in the spray core. For feed solid content of 40 wt.%, spray temperature varied from 52°C to 315°C, for solid content of 50 wt.%, flame temperatures are the highest, from 97°C to 515°C.

Increase of solid content in the feed results in higher viscosity of the solution, which reduces atomization pressure. The effect of atomization pressure on flame temperature was discussed earlier – for lower atomization pressure, air entrance into the spray core is enhanced due to larger spacing between droplets in the spray. Additionally, when solid content in the feed increases, flame temperature and length increase due to lower water-to-fuel ratio in the droplets.

For solid content of 30, 40, and 50 wt.% in the fuel-free solution after addition of 35 wt.% of ethanol gives water-to-fuel ratio 1.30, 1.11, and 0.93

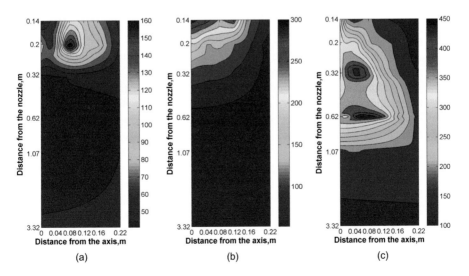

FIGURE 1.41 Experimental spray temperature pattern in the drier for different solid content in the fuel-free solution: (a) 30 wt.%, (b) 40 wt.%; (c) 50 wt.%. (Sobulska, Zbicinski, and Piatkowski 2020 Permission Granted.)

resp. Spray containing less fuel than water (water-to-fuel ratio 1.30 and 1.11) develops lower flame temperatures. In FSD test, where content of fuel is higher than content of water (water-to-fuel ratio 0.93), the highest flame temperature and flame length are observed (Figure 1.41c).

- Effect of fuel type

  The flame shape and temperature of spray are highly dependent on the volatility of the flammable component. To study the influence of the fuel type on temperature distributions and flame characteristics, Piatkowski, Taradaichenko, and Zbicinski (2014) dried ceramic powder using sunflower oil and ethanol at identical process parameters (atomization pressure of 0.8 MPa, nozzle type Fine 0.1, feed rate of 10 kg/h, fuel concentration of 50 wt.%, solid concentration of 50 wt.%, air flow rate of 200 Nm³/h) (Piatkowski, Taradaichenko, and Zbicinski 2014). The net heating values of sunflower oil and ethanol account for ca. 38 MJ/kg (Ryan, Dodge, and Callahan 1984) and 26.85 MJ/kg (Green and Perry 2008) resp.

  Comparison of the results displayed in Figure 1.42a and b shows differences in the combustion intensity for different volatility of the fuels (Yarin and Hetsroni 2004). Due to high volatility, ethanol evaporates immediately after atomization, which develops higher temperatures in the combustion zone, from 300°C to 700°C, and length of the flame from 0.5 to 1 m. The maximal temperature ca. 700°C observed at the spray core at the distance from 0.32 to 1.07 m from the nozzle.

  Lower temperatures in the atomization/flame zone (from 200°C to 500°C) and longer flame (from 1 to 1.5 m) can be observed for sunflower oil as a result of nonvolatile nature of the fuel. The maximal temperature of the flame 500°C was found at distance from 0.5 to 1.8 m from the nozzle in the spray core.

- Effect of air flow rate

  Sobulska, Zbicinski, and Piatkowski 2020 carried out FSD tests for air flow rate of 340 and 480 Nm³/h, different fuel and solid concentration in the feed (35 and 40 wt.% of solids in fuel free solution), constant feed rate of 5.5 kg/h, and atomization pressure of 0.7 MPa to determine the effect of airflow rate on FSD process. For air flow rate of 360 Nm³/h, spray temperatures varied from 101°C to 674°C (Figure 1.43a). The maximal flame temperatures of 510°C–674°C were found at the distances of 0.2 and 0.32 m from the nozzle. For higher air flow rate, 480 Nm³/h, flame temperatures were in the range from 83°C to 693°C (Figure 1.43b). Zone of maximal temperatures (657°C–693°C) covers larger area in the drying chamber compared with lower air flow rate. Higher air flow rate in the tower led to enhanced supply of air into the core of the spray and intensive fuel vapor/air mixing, generating higher temperatures in the spray core in the vicinity of nozzle. Air outlet temperature is slightly lower for higher air flow rate in tower: 101°C for test with air flow rate of 360 Nm³/h and 83°C for flow rate of 480 Nm³/h due to faster cooling of high-temperature flame by increased air supply and decreased overall equivalence ratio of excess of air in the air/fuel mixture.

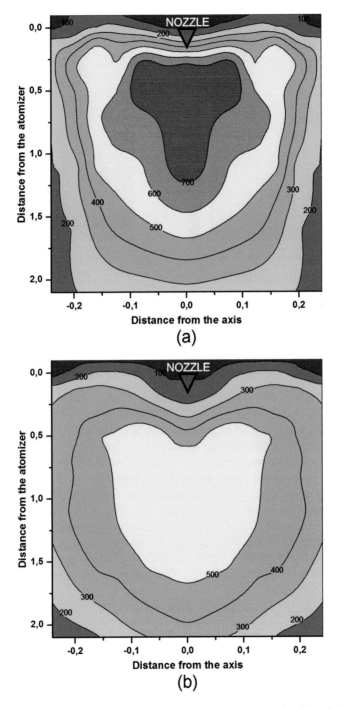

FIGURE 1.42 Temperature pattern in the drying zone due to combustion of (a) ethanol; (b) sunflower oil. (Piatkowski, Taradaichenko, and Zbicinski 2014 Permission Granted).

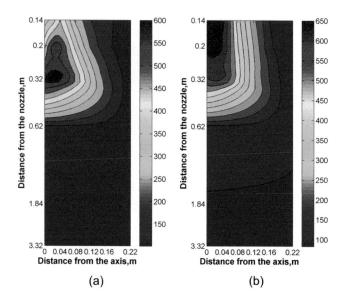

FIGURE 1.43 Experimental spray temperature pattern in the drier for different air flow rates in the tower: (a) 340 N m³/h, (b) 480 N m³/h. (Sobulska, Zbicinski, and Piatkowski 2020 Permission Granted.)

1.7 MATHEMATICAL MODELING OF FSD PROCESS

CFD models of different types of classical spray drying process have been already described in the literature (Langrish et al. 2020; Jaskulski, Wawrzyniak, and Zbiciński 2015; Poozesh, Lu, and Marsac 2018). Unlike SSD process, the modeling of flame spray drying requires modification of the Navier–Stokes equations such as implementation of additional mass, energy, and momentum conservation equations for the species taking part in the combustion reaction, modification of the energy equation to account for exothermic reaction and radiative heat transfer, selection of the proper turbulence model and phase change model. The discrete phase model (DPM) must be employed to calculate particle trajectories and heat and mass transfer between disperse and continuous phase. The Lagrangian framework can be applied for particles tracking, which involves the separate calculations of continuous and discrete phases where each particle represents a sample of particles that follow an identical path.

1.7.1 Modeling of Continuous Phase

1.7.1.1 Conservation Equations

Standard continuity conservation equation, conservation equation of mass species, equation for momentum conservation, and equation for energy conservation employed in the model are presented below (Chung 2002).

Combustion process involves chemical reaction between different species; therefore, the equation for mass conservation for reactive flow includes mixture conservation equation and conservation equations for species.

Continuity conservation equation can be written as:

$$\frac{\partial \rho}{\partial t} + \frac{\partial \rho v_i}{\partial x_i} = 0 \qquad (1.3)$$

The equation of mass conservation for mass fraction Y of component k with the addition of source term ω_k due to chemical reaction (Chung 2002):

$$\rho \frac{\partial Y_k}{\partial t} + \frac{\partial}{\partial x_i}\left(\rho Y_k v_i\right) - \frac{\partial}{\partial x_i}\left(\rho D_{km} Y_{k,i}\right) = \omega_k \qquad (1.4)$$

For reactive flows, the rate of change of the momentum is affected by body force acting on species k (Chung 2002).

$$\frac{\partial}{\partial t}\left(\rho v_j\right) + \frac{\partial}{\partial x_i}\left(\rho v_i v_j\right) + \frac{\partial}{\partial x_i}\left[p\partial_{ij} - \mu\left(\frac{\partial v_j}{\partial x_i} + \frac{1}{3}\frac{\partial v_i}{\partial x_j}\right)\right] = \rho \sum_{k=1}^{N} Y_k f_{kj} \qquad (1.5)$$

Where f_k is the external force per unit mass on species k, μ – viscosity, δ_{ij} is the Kronecker delta.

Equation of energy conservation includes source term for exothermic reaction (Chung 2002):

$$\frac{\partial}{\partial t}\left(\rho E\right) + \frac{\partial}{\partial x_i}\left(\rho E v_i + p v_i\right) - \frac{\partial}{\partial x_i}\left(kT_{,i} + \sum_k \rho HD_{km} Y_{k,i} + \tau_{ij} v_j\right) = W - \sum_k H_k^0 \omega_k \qquad (1.6)$$

1.7.1.2 Turbulence Model

Turbulent flows are characterized by the presence of oscillating velocity fields, which generates oscillations of transported quantities such as momentum, energy, and species concentration.

For simulation of combustion process, the standard k-ε model is frequently used (El Tahry 1983). In concurrent FSD system, the airflow pattern is uniform and without swirl. As the turbulence field could be assumed as isotropic, according to literature suggestions, the k-ε turbulence model is the most suitable model to predict continuous-phase turbulence in concurrent FSD.

The k-ε model is an empirical two-equation model, which includes two extra transport equations for the turbulence kinetic energy (k) and the specific dissipation rate (ε) in order to represent turbulence flow. The turbulence energy is determined by the first variable, i.e., turbulence kinetic energy, the second variable (ε) determines the scale of the turbulence.

The turbulent kinetic energy k and rate of dissipation ε are obtained from the following transport equations (Launder and Sharma 1974):

$$\rho \frac{Dk}{Dt} = \frac{\partial}{\partial x}\left[\left(\mu + \frac{\mu_t}{\delta_k}\right)\frac{\partial k}{\partial x}\right] + G_k + G_b - \rho\varepsilon - Y_M + S_k \qquad (1.7)$$

$$\rho \frac{D\varepsilon}{Dt} = \frac{\partial}{\partial x}\left[\left(\mu + \frac{\mu_t}{\delta_k}\right)\frac{\partial \varepsilon}{\partial x}\right] + C_{1\varepsilon}\frac{\varepsilon}{k}(G_k + C_{3\varepsilon}G_b) - C_{2\varepsilon}\rho\frac{\varepsilon^2}{k} + S_\varepsilon \qquad (1.8)$$

In equations (1.7) and (1.8), G_k represents the generation of turbulence kinetic energy due to the mean velocity gradients. G_b is the generation of turbulence kinetic energy due to buoyancy. Y_M represents the contribution of the fluctuating dilatation in compressible turbulence to the overall dissipation rate. $C_{1\varepsilon}$, $C_{2\varepsilon}$, and $C_{3\varepsilon}$ are constants, δ_k and δ_ε are the turbulent Prandtl numbers for k and ε, respectively, S_k and S_ε are user-defined source terms.

The turbulent (or eddy) viscosity μ_t is computed by combining k and ε as follows:

$$\mu_t = \rho \cdot C_\mu \frac{k^2}{\varepsilon} \qquad (1.9)$$

where C_μ is a constant.

The model constants $C_{1\varepsilon}, C_{2\varepsilon}, C_{3\varepsilon}, \sigma_k$, and δ_ε have the following default values after (Launder and Spalding 1972):

$$C_{1\varepsilon} = 1.44, \quad C_{2\varepsilon} = 1.92, \quad C_{3\varepsilon} = 0.09, \quad \sigma_k = 1.0, \quad \sigma_\varepsilon = 1.3 \qquad (1.10)$$

1.7.1.3   Implementation of the Combustion Model

The eddy dissipation model (EDM) is based on the assumption that chemical reaction is faster than diffusion in the flow ("ANSYS CFX-Solver Theory Guide" 2012). The model assumes that the reaction rate may be related directly to the time required to mix reactants at the molecular level. In turbulent flows, this mixing time is dominated by the eddy properties, and therefore, the combustion reaction rate is proportional to a mixing time defined by the turbulent kinetic energy, k, and dissipation, ε.

$$\text{Rate} \approx \frac{\varepsilon}{k} \qquad (1.11)$$

The EDM has been applied in many industrial combustion problems where reaction rates are faster than reactant mixing rates.

In the EDM, the rate of progress of elementary reaction i is determined by the smallest of the two following phenomena ("ANSYS CFX-Solver Theory Guide" 2012):

- reactants limiter

$$R_i = A\frac{\varepsilon}{k}\min\left(\frac{[C_k]}{v'_{ki}}\right) \qquad (1.12)$$

where A is EDM coefficient (equal to 4 by default), C_k is the molar concentration of component k, which only includes the reactant components, and v'_{ki} is the stoichiometric coefficient of the species k for reaction step i. R_i is elementary reaction rate for reaction i.

- products limiter

$$R_i = AB \frac{\varepsilon}{k} \left(\frac{\sum_k p(C_k) W_k}{\sum p v''_{ki} W_k} \right) \qquad (1.13)$$

where B is EDM coefficient, W_k is molar mass of the component k including only products of the reaction.

- maximum flame temperature limiter

A maximum flame temperature limiter may also be applied for the EDM. The reaction rate is smoothly aligned to zero when the specified upper temperature limit is approached. This is implemented by an additional limit added to the minimum condition in the EDM reaction rate:

$$R_{i,\mathrm{MFT}} = A \frac{\varepsilon}{k} C_{\mathrm{MFT}} \qquad (1.14)$$

where:

$$C_{\mathrm{MFT}} = \max \left\{ (T_{\max} - T), 0[K] \right\} \cdot \frac{\rho C_P}{\Delta H_R} \qquad (1.15)$$

C_{MFT} is a virtual concentration, which vanishes if the temperature is equal to the maximum flame temperature. C_P is the specific heat capacity of the fluid mixture at constant pressure, and ΔH_R is the reaction heat release per mole.

During elaboration of CFD model of FSD process, all described limiters EDM were tested; however, best agreement with experimental data was achieved for maximum flame temperature limiter to calculate the rate of combustion reaction.

1.7.1.4 Rate of Chemical Reaction

Chemical reactions are described by the chemical reaction equations in the form:

$$\sum_{k=1}^{N} v'_{ki} M_k \xrightarrow{k_f} \sum_{k=1}^{N} v''_{ki} M_k; (i = 1, \dots M) \qquad (1.16)$$

Where v_{ki} is the stoichiometric coefficient of the species k for the reaction step i, with the prime and double primes representing the reactant and product, resp. M_k is the chemical symbol for the component k. The specific reaction rates of the forward and backward reactions are k_f and k_b. Chemical reactions are governed by law of mass action, which implies that rate of chemical reaction is proportional to concentration of reactants ("ANSYS CFX-Solver Theory Guide" 2012):

$$\omega_k = W_k \sum_{i=1}^{M} (v''_{ki} - v'_{ki}) R_i \qquad (1.17)$$

where W_k is molar mass of component k. In CFD model presented in this work, R_i was determined using EDM.

In FSD process, ethanol vapor reacts with oxygen in the combustion reaction assuming a stoichiometric mixture of reagents (complete burning):

$$C_2H_6O + 3\,O_2 \xrightarrow{\;k_f\;} 2\,CO_2 + 3H_2O + 1{,}760\,kJ \tag{1.18}$$

This reaction can be classified as nonpremixed combustion because gaseous ethanol has to mix with oxygen due to molecular transport (diffusion) prior to reaction. In this case, chemical reaction is infinitely faster than components diffusion; therefore, we can assume that ethanol combustion is a fast one-step reaction (Dongmo et al. 2009).

One-step global Arrhenius rates, $R_{k,\,r}$, used for gas-phase reactions were computed as ("ANSYS CFX-Solver Theory Guide" 2012):

$$R_{k,r} = \left(v'_{k,r} - v''_{k,r}\right)\left(k_r \prod_j C_j^{\left(\eta'_{j,r}+\eta''_{j,r}\right)} \right) \tag{1.19}$$

$$k_r = A_r e^{Ea/R_k T} \tag{1.20}$$

where C_j is the molar concentration (kmol/m³) of species j in reaction i, $\eta'_{j,r}$ is the rate exponent for reactant species j in reaction r while $\eta''_{j,r}$ is the rate exponent of product species. Similarly, $v'_{k,r}$ is the stoichiometric coefficient for reactant species k in reaction r while $v''_{k,r}$ is that of product species. A_r is the pre-exponential factor, and E_a the activation energy (J/mol). Employed values for ethanol combustion reaction rates were as follows (Gröhn, Pratsinis, and Wegner 2012):

- pre-exponential factor of ethanol combustion reaction, $A_r = 8.435 \times 10^9$,
- activation energy for ethanol combustion reaction, $E_a = 1.256 \times 10^5$,
- rate exponent for ethanol, $\eta_{C2H6O} = 0.15$, stoichiometric coefficient for ethanol, $v_{C2H6O} = 1$,
- rate exponent for oxygen, $\eta_{O2} = 1.6$, stoichiometric coefficient for oxygen, $v_{O2} = 3$,
- rate exponent for carbon dioxide, $\eta_{CO2} = 0$, stoichiometric coefficient for carbon dioxide, $v_{CO2} = 2$,
- rate exponent for water, $\eta_{H2O} = 0$, stoichiometric coefficient for water, $v_{C2H6O} = 3$.

1.7.1.5 Thermal Radiation Model

The goal of radiation modeling is to solve the radiation transport equation, obtain the source term, W, for the energy equation, and the radiative heat flux at walls.

The spectral radiative transfer equation can be written as ("ANSYS CFX-Solver Theory Guide" 2012):

$$\frac{dI_V(\tau,s)}{dx} = \left(-(K_{av}+K_{sv})I_V(\tau,s)+K_{av}I_b(v,T)\right.$$

$$\left.+\frac{K_{sv}}{4\pi}\int_{4\pi} dI_V(\tau,s')\Phi(s,s')d\Omega'+I\right) \qquad (1.21)$$

where

v is frequency, τ – position vector, s – direction vector, x – path length, K_{av} – absorption coefficient, K_{sv} – scattering coefficient, I_b – blackbody emission intensity, I_V – spectral radiation intensity, which depends on position (r) and direction (s), T – local absolute temperature, Ω – solid angle, Φ – in-scattering phase function, I – radiation intensity source term.

Sobulska, Zbicinski, and Piatkowski (2020) applied the differential approximation or P1 thermal radiation model, which simplifies the radiation transport equation assuming that the radiation intensity is isotropic or direction is independent at a given location in space.

The radiative heat flux for an emitting, absorbing, and linearly scattering medium may be calculated as ("ANSYS CFX-Solver Theory Guide" 2012):

$$q_{rv} = -\frac{1}{3(K_{av}-K_{sv})-A_{ac}K_{sv}}\nabla G_v \qquad (1.22)$$

Substituting the above terms into the radiation transport equation, we obtain the equation for the spectral incident radiation ("ANSYS CFX-Solver Theory Guide" 2012):

$$-\nabla\cdot\left(\frac{1}{3(K_{av}-K_{sv})-A_{ac}K_{sv}}\nabla G_v\right) = K_{av}(G_v-4E_{bv}) \qquad (1.23)$$

1.7.2 Discrete-Phase Modeling

1.7.2.1 Liquid Evaporation Model

During FSD process, two liquids evaporate from the droplets – ethanol and water. The model uses two mass transfer correlations depending on whether the droplet is above or below the boiling point.

The boiling point is determined through the Antoine equation, i.e., vapor pressure equation describes the relation between vapor pressure and temperature for pure components ("ANSYS CFX-Solver Theory Guide" 2012).

$$p_{vap} = p_{scale}\,\exp\left(A_{ant}-\frac{B_{ant}}{T_p+C_{ant}}\right) \qquad (1.24)$$

where A_{ant}, B_{ant}, and C_{ant} are coefficients, which for ethanol and water have following values:

- for water $A_{ant}=4.6543$, $B_{ant}=1{,}435.264$, $C_{ant}=-64.848$,
- for ethanol $A_{ant}=5.24677$, $B_{ant}=1{,}598.673$, $C_{ant}=-46.424$(Yaws 2009).

When the particle is above the boiling point, the mass transfer is determined by ("ANSYS CFX-Solver Theory Guide" 2012):

$$\frac{dm_p}{dt} = -\frac{\pi \cdot d\lambda Nu(T_G - T_P) + \varepsilon_p \pi \cdot d_p^2 \left(I_p - \sigma \cdot n_f{}^2 T_p^4\right)}{H_v} \quad (1.25)$$

When the particle is below the boiling point, the mass transfer is given by:

$$\frac{dm_p}{dt} = -\pi \cdot d_p \rho DSh \frac{W_C}{W_G} \ln\left(\frac{1 - X_s^v}{1 - X_{vap}^v}\right) \quad (1.26)$$

Where d_p is the droplet diameter, ρD is the dynamic diffusivity of the component in the continuum, and Sh is the Sherwood number. W_C and W_G are the molecular weights of the vapor and the mixture in the continuous phase, X_s^v is the equilibrium vapor mole fraction of the evaporating component at the droplet surface, and X_{vap}^v is the mole fraction of the evaporating component in the gas phase.

The mass source to the continuous fluid is obtained from:

$$\frac{dS}{dt} = -\frac{dm_p}{dt} \quad (1.27)$$

The standard energy balance describes heat transfer between particles and the surrounding air:

$$\frac{d\left(m_p C_p T_p\right)}{dt} = \pi d_p \lambda Nu(T_G - T_P) + \frac{dm_p}{dt} V \quad (1.28)$$

The mass and energy transfer between droplets and drying air are taken into account, where the Ranz–Marshall correlations were used to calculate Nusselt and Sherwood numbers:

$$Nu = 2 + 0.6\,\mathrm{Re}^{\frac{1}{2}} \mathrm{Sc}^{\frac{1}{3}} \quad (1.29)$$

$$Sh = 2 + 0.6\,\mathrm{Re}^{\frac{1}{2}} \mathrm{Pr}^{\frac{1}{3}} \quad (1.30)$$

1.7.3 COMPUTATIONAL MESH AND INITIAL PSD

Sobulska, Zbicinski, and Piatkowski (2020) carried out CFD steady-state simulations of FSD in 3D geometry of experimental TUL spray drying tower (Figure 1.3). In FSD CFD model, high-quality computational mesh of 82,000 tetrahedral elements was used (Figure 1.44). The following mesh quality parameters were reported: minimum orthogonal quality was 0.42, maximum aspect ratio was 9.64. Orthogonal quality close to 0 corresponds to low mesh quality and orthogonal quality close to 1 indicates high quality of the mesh, the acceptable minimum orthogonal quality for cells is more than 0.01. The aspect ratio equal to 1 defines the highest mesh quality.

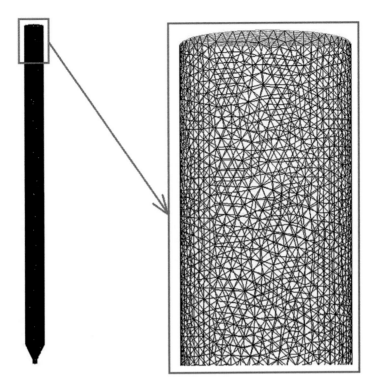

FIGURE 1.44 Computational mesh applied in CFD model.

Converged solution results obtained for continuous phase were applied for initialization of calculations of full coupling model between continuous and discrete phase (flow of continuous phase with injected particles). To obtain converged solutions, the residuals of all the variables must have achieved values below 10^{-5}.

Increase of particle diameter along the drying tower height caused by particle puffing and agglomeration was not considered in the FSD CFD model.

Initial particle size and velocity distributions obtained experimentally at the distance of 0.025 m from the nozzle were used as boundary conditions for particle injection. Rosin–Rammler function of the PSD (Rosin and Rammler 1933) was determined for each position using PDA measurements performed in 50 positions along the drying tower.

FSD CFD model was solved introducing 50 particle injection regions using Rosin–Rammler functions for PSD and average particle velocities determined experimentally. Figure 1.45 shows particle injection path lines in ANSYS CFX software colorized by particle velocities. Figure 1.46 illustrates experimental average particle velocities at different points along the drying tower diameter. The figures show that particle velocity distribution is symmetrical regarding drying chamber axis. Droplet's velocity decreases along the drying tower radius; highest droplet velocity 17 m/s was found at the tower axis, whereas at the spray edge, droplet velocity was equal to 2.5 m/s.

FIGURE 1.45 Particles injection points.

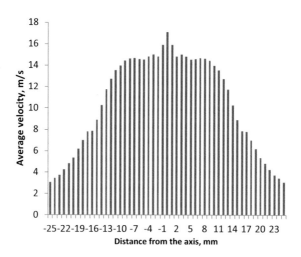

FIGURE 1.46 Initial particle velocities determined experimentally: average velocities at different positions at tower cross section.

1.7.4   RESULTS OF FSD SIMULATIONS

1.7.4.1   Air Temperature Distribution

Figures 1.47–1.50 show the average, maximal, and minimal temperature distributions in the tower cross section for different distances from the nozzle: 0.14, 0.20, 0.32, and 0.62 m determined experimentally and theoretically.

Due to temperature fluctuations during FSD process, only the average experimental spray temperature can be compared with theoretically calculated. Red dashed lines represent minimal and maximal temperatures registered during 2 min of measurement, and red spots show the average temperatures.

At the distance from the nozzle of 0.14 m (Figure 1.47), the maximal flame temperature fluctuations were found at the spray core where the highest flame temperature archived values 1,006°C–1,170°C and minimal flame temperature varied from 63°C to 80°C. At the edges of the spray, out of the flame zone, temperature fluctuations are smaller, at point 0.22 m from the axis maximal flame temperature is 238°C, minimal 43°C.

Similar analysis at the distance of 0.2 m from the nozzle (Figure 1.48) shows large flame temperature oscillations in the flame core; maximal flame temperature is in range of 814°C–1,019°C and minimal from 123°C to 82°C. At the spray edge (position 0.22 m from the axis), flame temperature changes between 196°C and 42°C.

At distance of 0.32 m from the nozzle (Figure 1.49), maximal flame temperatures oscillate from 1,193°C to 1,721°C at the spray core. At the spray edge, maximal flame temperature decreased to 320°C. The highest flame temperatures were registered at level 0.32 m in the region just above the flame end.

At distance of 0.62 m from the nozzle (Figure 1.50), maximal flame temperature is in the range from 890°C to 1,120°C.

In the analyzed FSD test, flame length varies from 0.32 to 0.62 m, and combustion process is unstable; therefore, some discrepancies are observed between the results of the experiments and simulation at distance of 0.32 m, positions of 0.12 and 0.16 m from the nozzle (Figure 1.49) and at distance of 0.62 m, positions of 0.04 and 0.08 m (Figure 1.50).

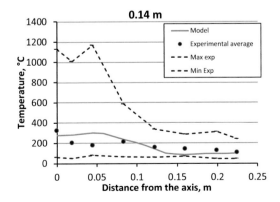

FIGURE 1.47   Experimental and theoretical air temperature distributions at the distance of 0.14 m from the nozzle. (Sobulska, Zbicinski, and Piatkowski 2020 Permission Granted.)

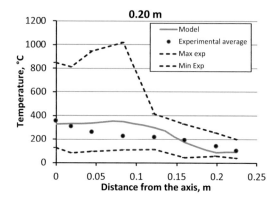

FIGURE 1.48 Experimental and theoretical air temperature distributions at the distance of 0.20 m from the nozzle. (Sobulska, Zbicinski, and Piatkowski 2020 Permission Granted.)

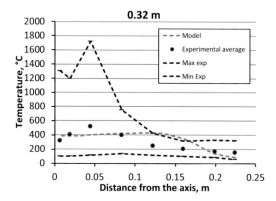

FIGURE 1.49 Experimental and theoretical air temperature distributions at the distance of 0.32 m from the nozzle. (Sobulska, Zbicinski, and Piatkowski 2020 Permission Granted.)

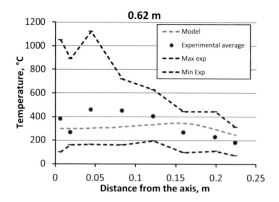

FIGURE 1.50 Experimental and theoretical air temperature distributions at the distance of 0.62 m from the nozzle. (Sobulska, Zbicinski, and Piatkowski 2020 Permission Granted).

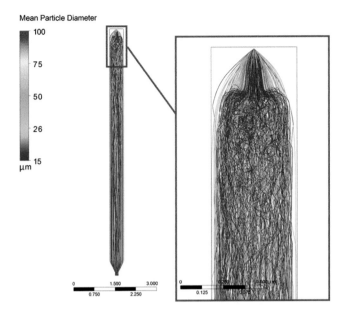

FIGURE 1.51   Particle path lines colored by mean particle diameter.

Figure 1.51 shows particle paths colored by particle diameters and in the vicinity of particle injection points. At the distance ca. 0.15 m from the particle's injection, the upward flow and intensive mixing of the particles caused by flow turbulence due to combustion process are observed. At the distance of 1.5 m from the injection point, particle path lines became parallel to the tower axis, which is in line with particle flow pattern observed in PDA experiments.

1.7.4.2   Combustion Process

CFD model allows analysis of reactants and products concentrations of chemical reaction of combustion taking place during FSD. Figures 1.52 and 1.53 show concentrations of the ethanol and oxygen taking part in the chemical reaction of combustion for two drying tests carried out for different feed solids content in the fuel-free solution of 40 and 50 wt.%, feed rate of 10 kg/h, fuel concentration of 35 wt.%. Figure 1.52 presents the mass fraction of ethanol vapor in the upper part of the drying tower. The highest mass fraction of ethanol vapor (ca. 0.096) is observed in the core of the spray indicating the zone of fuel evaporation, which is in agreement with suggestions of (Świątkowski 1998), which distinguished four flame zones developed during liquid spray combustion: zone of initial heating and evaporation, zone of fuel vaporization, and zone of combustion, where combustion products are generated. The stream of atomized fuel is mixed with the air and heated up by the hot air and combustion products coming from the reaction zone. Droplets evaporation is fast on the side edge of the spray and slow in the spray core. The ignition starts if temperature of the mixture of fuel vapor and air is high enough to initiate the combustion. Combustion takes place in the outer layer of the stream, which has the highest temperature. In the core of the stream, the droplet temperatures are too low to provoke the ignition of the

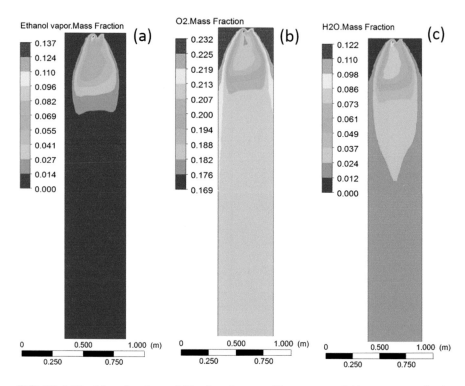

FIGURE 1.52 Mass fractions of (a) ethanol vapor, (b) oxygen, and (c) water vapor in the drying tower (solid content 50 wt.%).

stream due to cooling effect of droplet evaporation. Results of mathematical modeling show that in FSD test with solid content of 50 wt.%, the concentration of ethanol vapor decreases near the spray edges to ca. 0.027 and then to 0.0 as ethanol reacts with oxygen in the combustion reaction.

Figure 1.52b shows the concentration of oxygen in the drying tower during FSD process. At the drying tower inlet, the mass fraction of oxygen is 0.23, which corresponds to the oxygen content in air. As oxygen reacts with ethanol vapor, the mass fraction of oxygen decreases to ca. 0.180 in the flame core. Below the flame mass fraction of oxygen is constant and equal to ca. 0.210.

Figure 1.52c shows the mass fractions of water vapor in the upper part of the drying tower. During FSD process, water vapor is produces in two routes, i.e., by evaporation from the droplets and as a product of chemical reaction of combustion. As a result, the highest amount of water vapor is observed in the spray core, ca. 0.086. At the spray edges, mass fraction of water vapor is reduced to 0.024 and out of the spray region is equal to mass fraction of water vapor in the inlet air of <0.012.

Figure 1.53 shows the mass fraction of chemical species for drying test, where content of solids in the feed is lower, i.e., 40 wt.%. Mass fraction of ethanol vapor changes from 0.017 in the spray core to 0.0 at the spray edges (Figure 1.53a). The length of ethanol evaporation area in FSD test with solid content of 40 wt.% is ca.

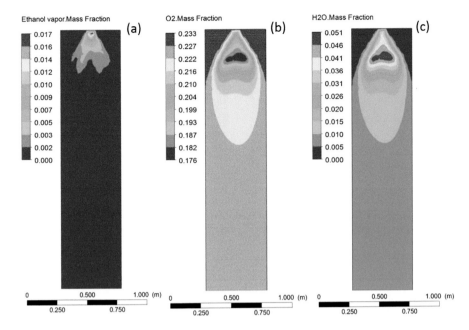

FIGURE 1.53 Mass fractions of (a) ethanol vapor, (b) oxygen, and (c) water vapor in the drying tower (solid content 40 wt.%).

0.25 m, whereas in test with solid content of 50 wt.% – ca. 0.65 m, which may be explained by the differences in flame length: 0.2 and 0.62 m for both tests, resp.

Figure 1.53b shows mass fraction of oxygen in the drying tower. The maximum oxygen concentration of 0.23 is observed at the drying air inlet and at the outward regions of the spray. The minimum oxygen mass fraction ca. 0.17 was found at the spray core. The water vapor concentration is the highest (0.051) at the spray core where water vapor occurs due intensive moisture evaporation and combustion reaction.

1.7.4.3 Particle Drying Time

CFD model of the FSD process allows to determine particle drying time. Figure 1.54 shows drying curves for particles with two different initial diameters of 33 and 53 μm. Drying time for smaller and larger particles is 0.06 and 0.12 s resp.

According to the experimental data, the average particle residence time in flame for test with solid content of 50 wt.% is 0.214 s. Recalculating particle drying time to the particle traveling distance, one can find that small particles (33 μm) are completely dried at the distance of 0.33 m from the nozzle, whereas large particles (53 μm) at 0.45 m (Figure 1.54).

Analysis of the temperature distribution at distances of 0.33 and 0.45 m presented in the Figure 1.41c shows that for FSD test with solid content of 50 wt.%, the particles are dried faster than they left the high-temperature flame region.

Short particle drying time compared with particle residence time in flame explains the high values of color index (84.9) and high content of caramelization products

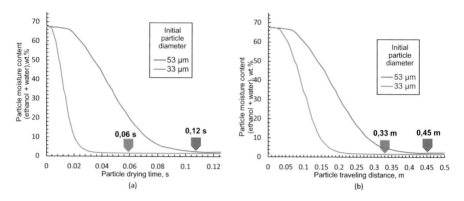

FIGURE 1.54 Particle drying time and traveling distance calculated from CFD model (solid content 50 wt.%). (Sobulska, Zbicinski, and Piatkowski 2020 Permission Granted).

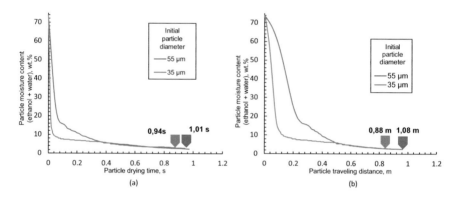

FIGURE 1.55 Particle drying time and traveling distance calculated from CFD model (solid content 40 wt.%). (Sobulska, Zbicinski, and Piatkowski 2020 Permission Granted.)

(228.3 mg/kg) in the dried powder for test with high solid content indicating particles overheating.

Particle drying times determined from CFD model of FSD for test with lower content of solid in the feed – 40 wt.% is shown in Figure 1.55. Drying time for particle with initial diameter 35 µm was 0.94 s, and for particle with higher initial diameter of 55 µm, drying time is 1.01 s. Particle residence time in flame determined experimentally is 0.03 s for test with solid content of 40 wt.%. Figure 1.55 shows the distance that the particles covered during drying: 0.88 m for small particles and 1.08 m for large particles, in this area, spray temperature is ca. 65°C (Figure 1.41a). Due to longer drying time compared with residence time in flame, we observe enhanced particle quality for FSD test with low solid content: lower color index is 18.7 and lower HMF content is 116.2 mg/kg, compared with tests with solid content of 50 wt.%.

Final conclusion is that product quality in FSD process might be controlled by optimal selection process parameters.

1.8   SCALE-UP RULES AND PROCEDURE

Direct scaling-up of the FSD process from pilot to industrial scale via increase of feed flow rate or feed concentration is limited due to the following obstacles:

- Increase of feed flow rate (without change of size of nozzle orifice) affects the initial droplet size distribution, which is related with flame temperature and length as well as particle residence time in flame. Therefore, direct increase of feed flow rate will impact the product properties such as particle size distribution and affects product thermal degradation, which are determined by particle residence time in flame.
- Direct increase of feed concentration in FSD process also results in substantial increase of flame temperature and length. For instance, the highest thermal degradation of the product is observed for FSD tests with highest solid concentrations in the feed.

Since the main constraint during scaling-up is repeatability of final product properties such as PSD, bulk density, thermal degradation, following general scaling-up rules of FSD process are proposed:

- Increase of feed rate should be coupled with change of nozzle type with possible increase of nozzle orifice in order to keep the same atomization pressure and initial droplet size distribution of the spray as in laboratory-scale tests.
- Increase of solid content in the feed should be carried out with simultaneous increase of air flow rate in the tower to keep the same temperature of the drying medium at the dryer outlet as in laboratory tests.

To decrease the cost of development and construction of flame spray dryers at the industrial-scale development of reliable mathematical CFD model, which accounts for the detailed description of combustion chemistry, complex fluid, and particle dynamics including particle formation mechanisms such as agglomeration and puffing can be crucial.

The mathematical model presented in Section 1.7 is the first attempt for simulation of FSD process but was developed with a number of assumptions and simplifications, e.g., steady-state conditions; therefore, development of advanced mathematical models of simultaneous combustion and drying of droplets is still required to enable design and scaling-up of the FSD dryers. Further development of FSD CFD model could be implementation of detailed description of droplet drying kinetics including both first and second drying stages (present model assumes pure water droplet drying without decrease of drying rate after solid crust formation).

The application of CFD modeling of FSD process as an effective tool for design and scaling-up of the dryers is still limited due to the following challenges:

- Description of the combustion chemistry: majority of the CFD models applied for simulation of flame spray processes as FSD, pulse combustion drying (PCD), FSP, and emulsion combustion method (ECM) assume single or two-step chemical reaction of fuel combustion; however, the real combustion process may include large amount of species and numerous chemical reactions.

- Lack of understanding of the complex combustion mechanisms.
- Insufficient accuracy of turbulence models for combustion processes.
- Difficulty with detailed descriptions of droplet drying kinetics and particle formation mechanism via agglomeration and puffing due to rapid evaporation and the presence of the recirculation zones in the drying chamber.

1.9 SUMMARY

Theoretical and experimental analyses of FSD process show that increase of solid content in the feed, decrease of feed rate, and lower atomization pressure result in larger droplets in the spray, smaller spraying angle, which produce longer flame, higher temperatures in the combustion zone, and longer particle residence time in the flame.

For fine atomization, the combustion takes place only in the outward regions of the spray producing short flame due to limited entrance of the air into the core of the spray, and vice versa; for coarse atomization, enhanced entrance of the air into the spray core results in intensive combustion and longer flame.

Growth of particle diameters of skin-forming materials (e.g., maltodextrin) occurs through two routes: particle agglomeration, which takes place in the recirculation zones formed at the side edges of the flame and puffing/inflation, which depends on the particle residence time and temperature in the combustion zone.

Relationship between particle residence time in flame and drying time is key parameter affecting the quality of the dried product, for drying time shorter than particle residence time in the flame, color index and HMF content are substantially higher than for short particle residence time in the flame.

NOMENCLATURE

A = EDM coefficient (equal to 4)
A_{ac} = linear anisotropy coefficient
A_{ant}, B_{ant} and C_{ant} = coefficients in the Antoine equation
A_r = pre-exponential factor in Arrhenius equation (s^{-1})
B = EDM coefficient (equal to 1)
B_m = mass transfer number
B_t = heat transfer number
C = molar concentration (mol/m^3)
$C_{1\varepsilon}, C_{2\varepsilon}, C_{3\varepsilon}, \delta_k, \delta_\varepsilon$ = turbulent constants in equations (6.5), (6.6) and (6.8)
C_{MFT} = virtual concentration (mol/m^3) in equation (6.13)
C_p = heat capacity (J/kg K^{-1})
D or d = particle diameter (m)
D_g = diffusion coefficient of vapor (m^2/s)
D_{km} = diffusion coefficient of component k, (m^2/s)
dm_p/dt = rate of evaporation (kg/s)

dS/dt = mass added to the continuous phase from the discrete phase (kg/s)

E = energy (J/s)

E_a = the activation energy (J/mol)

E_{bv} = energy spectrum for radiation emitted by a blackbody (J/s)

f_k = the external force per unit mass on species k

G = group combustion number

G_b = the generation of turbulence kinetic energy due to buoyancy (J/s)

G_k = the generation of turbulence kinetic energy due to the mean velocity gradients (J/s)

Gr = Grashof number

G_v = spectral incident radiation (W/m^2)

H_V = latent heat of evaporation (J/kg)

ΔH or H = heat of combustion (J/kg)

I = radiation intensity source term (W/m^2)

I_b = blackbody emission intensity (W/m^2)

I_p = the radiation intensity on the particle surface (W/m^2)

I_V = spectral radiation intensity, which depends on position (r) and direction (s) (W/m^2)

k = turbulence kinetic energy (m^2/s^3)

K_{av} = absorption coefficient (m^{-1})

K_{sv} = scattering coefficient (m^{-1})

l = spacing between droplets (m)

Le = Lewis number

$\dot{m}$ = mass flow rate (kg/s)

m = mass (kg)

M_k = the chemical symbol for the component k

n = number of droplets or particles

N_d = number of droplets in the cloud

n_f = the refractive index of the fluid

Nu = Nusselt number

p = static pressure (Pa)

Pr = Prandtl number $\left(c_p \mu / k_\infty \right)$

$\dot{Q}$ = heat flux (W)

Q_{vap} = heat for moisture evaporation (MJ/h)

r = radius (m)

R = universal gas constant

Re = Reynolds number

R_i = rate of progress of elementary reaction i (mol/L/s)

s = direction vector

S = particle traveling distance (m)

Sc = Schmidt number, $\left(\dfrac{\mu}{\rho D_{i,m}} \right)$

Sh = Sherwood number

T = temperature (K)

t = time (s)

u_i = velocity of particular fraction (m/s)
U_w = average-weighted velocity of particles fractions (m/s)
v_i = axial velocity (m/s)
v_j = radial velocity (m/s)
W = energy source term (W)
W_C = molar mass of the mixture in the continuous phase (kg/mol)
W_G = molar mass of the vapor (kg/mol)
W_k = molar mass of the component k (kg/mol)
X = moisture content (kg$_{H2O}$/kg$_{dry\ material}$).
X' = moisture content (%)
x = path length (m)
x_i = the axial coordinate
x_j = the radial coordinate
X_S^v = the equilibrium vapor mole fraction of the evaporating component at the droplet surface
X_{vap}^v is the mole fraction of the evaporating component in the gas phase
Y = mass fraction
Φ = in-scattering phase function
Ω = solid angle (°)

GREEK SYMBOLS

∇ = Laplace operator
ϕ = equivalence ratio
δ_{ij} = Kronecker delta
ε_p = emissivity of particle
$_\varepsilon$ = turbulent kinetic energy dissipation (m²/s³)
$\eta''_{j,r}$ = the rate exponent of product species
$\eta'_{j,r}$ = the rate exponent for reactant species j in reaction r
λ = thermal conductivity (W/mK⁻¹)
μ = viscosity of the fluid (kg/ms)
ρ_b = bulk density (kg/m³)
ρ_2, ρ_1 and ρ_s = particle, liquid, and solid densities (kg/m³)
σ = the Stefan–Boltzmann constant (5.67×10^{-8} W·m⁻²·K⁻⁴)
τ = position vector
τ_{ij} = stress tensor
υ = frequency (1/s)
υ''_{ki} = stoichiometric coefficient of the product species k for reaction r
υ'_{ki} = stoichiometric coefficient of the reactant species k for reaction r
φ = coefficient in equations (2.6 and 2.7)
ω_k = mass source due to chemical reaction (kg/h)

ACRONYMS

AMD – Arithmetic Mean Diameter
CARS – Coherent Anti-Stokes-Raman

CFD – Computational Fluid Dynamics
DPM – Discrete-Phase Model
EDM – Eddy Dissipation Model
FSD – Flame Spray Drying
FSP – Flame Spray Pyrolysis
HMF – Hydroxymethylfurfural
LDA – Laser Doppler Anemometry
LIF – Laser-Induced Fluorescence
LIP – Laser-Induced Phosphorescence
PCD – Pulse Combustion Drying
PDA – Particle Dynamic Analysis
PSD – Particle Size Distribution
SMD – Sauter Mean Diameter
SSD – Standard Spray Drying

REFERENCES

"ANSYS CFX-Solver Theory Guide." 2012. http://www.ansys.com.

Arulprakasajothi, M., and P. L. Rupesh. 2020. "Surface Temperature Measurement of Gas Turbine Combustor Using Temperature-Indicating Paint." *International Journal of Ambient Energy*: 1–4. doi:10.1080/01430750.2020.1731709.

Buckmaster, J., and N. Peters. 1988. "The Infinite Candle and Its Stability-A Paradigm for Flickering Diffusion Flames." *Symposium (International) on Combustion* 21 (1): 1829–1836. doi:10.1016/S0082-0784(88)80417-X.

Char, J. M., and J. H. Yeh. 1996. "The Measurement of Open Propane Flame Temperature Using Infrared Technique." *Journal of Quantitative Spectroscopy and Radiative Transfer* 56 (1): 133–144. doi:10.1016/0022-4073(96)00013-1.

Chiu, H. H., and T. M. Liu. 1977. "Group Combustion of Liquid Droplets." *Combustion Science and Technology* 17 (3–4): 127–142. doi:10.1080/00102207708946823.

Chung, T. J. 2002. *Computational Fluid Dynamics*. Cambridge: Cambridge University Press. doi:10.1017/CBO9780511606205.

Dayton, D.C., and Th. D. Foust. 2020. "Optimized Biofuels for High-Efficiency, Low-Emission Engines." In *Analytical Methods for Biomass Characterization and Conversion*, 129–145. New York: Elsevier. doi:10.1016/b978-0-12-815605-6.00009-3.

Dongmo, E., R. Gadow, A. Killinger, and M. Wenzelburger. 2009. "Modeling of Combustion as Well as Heat, Mass, and Momentum Transfer during Thermal Spraying by HVOF and HVSFS." *Journal of Thermal Spray Technology* 18 (5–6): 896–908. doi:10.1007/s11666-009-9341-2.

El Tahry, Sh.H. 1983. "K- Epsilon Equation for Compressible Reciprocating Engine Flows." *Journal of Energy* 7 (4): 345–353. doi:10.2514/3.48086.

Eltz, H. U., F. Schoen, K. Hofmann, and H. Tappe. 1985. "Method for Thermosol Dying Polyester Fiber Material Using Liquid Dispersion Dye Preparations." German Patent DE3528261A1.

Filková, I., L. X. Huang, and A. S. Mujumdar. 2014. "Industrial Spray Drying Systems." In *Handbook of Industrial Drying*, 4th edition, edited by A.S. Mujumdar, 191–226. Boca Raton, FL: CRC Press. doi:10.1201/b17208.

García-Armingol, T., Á. Sobrino, E. Luciano, and J. Ballester. 2016. "Impact of Fuel Staging on Stability and Pollutant Emissions of Premixed Syngas Flames." *Fuel* 185: 122–132. doi:10.1016/j.fuel.2016.07.086.

Gounder, J. D., A. Kourmatzis, and A. R. Masri. 2012. "Turbulent Piloted Dilute Spray Flames: Flow Fields and Droplet Dynamics." *Combustion and Flame* 159 (11): 3372–3397. doi:10.1016/j.combustflame.2012.07.014.

Green, D. W., and R. H. Perry. 2008. *Perry's Chemical Engineers' Handbook*. New York: McGraw-Hill. https://www.accessengineeringlibrary.com/content/book/9780071422949.

Gröhn, A. J., S. E. Pratsinis, and K. Wegner. 2012. "Fluid-Particle Dynamics during Combustion Spray Aerosol Synthesis of ZrO2." *Chemical Engineering Journal* 191: 491–502. doi:10.1016/j.cej.2012.02.093.

Hahn, J., and T.F. Edgar. 2003. "Process Control Systems." In *Encyclopedia of Physical Science and Technology*, 111–126. New York: Elsevier. doi:10.1016/b0-12-227410-5/00612-8.

"Haynes International - Principal Features Hastelloy S." 2020. Accessed December 8. https://haynesintl.com/alloys/alloy-portfolio_/High-temperature-Alloys/hastelloy-s-alloy/principal-features.

"Haynes International - Principal Features Hastelloy X." 2020. Accessed December 8. https://www.haynesintl.com/alloys/alloy-portfolio_/High-temperature-Alloys/HASTELLOY-X-alloy/HASTELLOY-X-principal-features.aspx.

Heine, M. C., and S. E. Pratsinis. 2005. "Droplet and Particle Dynamics during Flame Spray Synthesis of Nanoparticles." *Industrial & Engineering Chemistry Research* 44 (16): 6222–6232.

Heitor, M. V., and L. N. Moreira. 1993. "Thermocouples and Sample Probes for Combustion Studies." *Progress in Energy and Combustion Science* 19: 259–278.

Huang, Y., and V. Yang. 2009. "Dynamics and Stability of Lean-Premixed Swirl-Stabilized Combustion." *Progress in Energy and Combustion Science* 35 (4): 293–364. doi:10.1016/j.pecs.2009.01.002.

Imaoka, R. T., and W. A. Sirignano. 2005. "Vaporization and Combustion in Three-Dimensional Droplet Arrays." *Proceedings of the Combustion Institute* 30 (2): 1981–1989. doi:10.1016/j.proci.2004.08.049.

"International Standard IEC 60584-1:2013. Thermocouples - Part 1: EMF Specifications and Tolerances." 2013.

Jaskulski, M., P. Wawrzyniak, and I. Zbiciński. 2015. "CFD Model of Particle Agglomeration in Spray Drying." *Drying Technology* 33 (15–16): 1971–1980. doi:10.1080/07373937.2015.1081605.

Jocher, A., J. Bonnety, T. Gomez, H. Pitsch, and G. Legros. 2019. "Magnetic Control of Flame Stability: Application to Oxygen-Enriched and Carbon Dioxide-Diluted Sooting Flames." *Proceedings of the Combustion Institute* 37 (4): 5637–5644. doi:10.1016/j.proci.2018.05.156.

JumoIndustries. 2014. "Industrial Furnace Construction." 1–39.

Kashir, B., S. Tabejamaat, and M. Baig Mohammadi. 2012. "Experimental Study on Propane/Oxygen and Natural Gas/Oxygen Laminar Diffusion Flames in Diluting and Preheating Conditions." *Thermal Science* 16 (4): 1043–1053. doi:10.2298/TSCI110524122K.

Kemp, I.C. 2011. "Fundamentals of Energy Analysis of Dryers." In *Modern Drying Technology. Energy Savings*, edited by E. Tsotsas and A. S. Mujumdar, 1–46. Hoboken, NJ: John Wiley and Sons Ltd.

Knoop, P., F. E. C. Culick, and E. E. Zukoski. 1997. "Extension of the Stability of Motions in a Combustion Chamber by Nonlinear Active Control Based on Hysteresis." *Combustion Science and Technology* 123 (1–6): 363–376. doi:10.1080/00102209708935635.

Kudra, T. 2008. "Pulse-Combustion Drying: Status and Potentials." *Drying Technology* 26 (12): 1409–1420. doi:10.1080/07373930802458812.

Langrish, T. A. G., J. Harrington, X. Huang, and Ch. Zhong. 2020. "Using CFD Simulations to Guide the Development of a New Spray Dryer Design." *Processes* 8 (932): 1–22.

Lapp, M., and C. M. Penney. 1979. "Instantaneous Measurements of Flame Temperature and Density by Laser Raman Scattering." In *Proceedings of the Dynamic Flow Conference 1978 on Dynamic Measurements in Unsteady Flows*, 665–683. doi:10.1007/978-94-009-9565-9_38.

Launder, B. E., and B. I. Sharma. 1974. "Application of the Energy-Dissipation Model of Turbulence to the Calculation of Flow near a Spinning Disc." *Letters in Heat and Mass Transfer* 1 (2): 131–137. doi:10.1016/0094-4548(74)90150-7.

Launder, B. E., and D. B. Spalding. 1972. *Lectures in Mathematical Models of Turbulence.* London, New York: Academic Press.

Law, C. L., and A. S. Mujumdar. 2010. "Energy Efficiency and Energy Savings in Drying." In *Drying of Foods, Vegetables and Fruits*, edited by S. V. Jangam, C. L. Law, and A. S. Mujumdar, 145–158. Singapore: National University of Singapore.

Lawitzki, A., I. Plath, W. Stricker, J. Bittner, U. Meier, and K. Kohse-Höinghaus. 1990. "Laser-Induced Fluorescence Determination of Flame Temperatures in Comparison with CARS Measurements." *Applied Physics B Photophysics and Laser Chemistry* 50 (6): 513–518. doi:10.1007/BF00408777.

Lefebvre, A. 1988. *Atomizations and Sprays*. Boca Raton, FL: CRC Press.

Leipertz, A., S. Pfadler, and R. Schießl. 2010. "An Overview of Combustion Diagnostics." In *Handbook of Combustion. Vol. 5:Combustion Diagnostics and Pollutants*, edited by M. Lackner, F. Winter, and A. Agarwal, 1–50. Weinheim, Germany: Wiley-VCH Verlag GmbH & Co. KGaA.

Marrocco, M. 2010. "CARS Spectroscopy." In *Handbook of Combustion. Combustion Diagnostics and Pollutants*, edited by M. Lackner, F. Winter, and A. K. Agarwal, 155–188. Weinheim, Germany: Wiley-VCH Verlag GmbH & Co. KGaA.

Min, J., and F. Baillot. 2012. "Experimental Investigation of the Flame Extinction Processes of Nonpremixed Methane Flames inside an Air Coflow Diluted with CO_2, N_2, or Ar." *Combustion and Flame* 159 (12): 3502–3517. doi:10.1016/j.combustflame.2012.05.015.

Mueller, R., L. Mädler, and S. E. Pratsinis. 2003. "Nanoparticle Synthesis at High Production Rates by Flame Spray Pyrolysis." *Chemical Engineering Science* 58 (10): 1969–1976. doi:10.1016/S0009-2509(03)00022-8.

"Nimonic 263 Nickel Based Alloy Supplier | HARALD PIHL | Harald Pihl." 2020. Accessed December 8. https://www.haraldpihl.com/en/products/nickel-alloys/nimonic-263/.

"Optris GmbH." Berlin, Germany. https://www.optris.com/.

Paschereit, Ch.O., and E. Gutmark. 2004. "The Effectiveness of Passive Combustion Control Methods." *Proceedings of the ASME Turbo Expo 2004*, 1: 361–374 doi:10.1115/gt2004-53587.

Piatkowski, M., M. Taradaichenko, and I. Zbicinski. 2014. "Flame Spray Drying." *Drying Technology* 32 (11): 1343–1351. doi:10.1080/07373937.2014.903413.

Piatkowski, M., M. Taradaichenko, and I. Zbicinski. 2015. "Energy Consumption and Product Quality Interactions in Flame Spray Drying." *Drying Technology* 33 (9): 1022–1028. doi :10.1080/07373937.2014.924137.

Piatkowski, M., and I. Zbicinski. 2013. "Spray Drying Method." Polish Patent 213458.

Poozesh, S., K. Lu, and P. J. Marsac. 2018. "On the Particle Formation in Spray Drying Process for Bio-Pharmaceutical Applications: Interrogating a New Model via Computational Fluid Dynamics." *International Journal of Heat and Mass Transfer* 122: 863–876. doi:10.1016/j.ijheatmasstransfer.2018.02.043.

Rayleigh, J. W. S. 1945. *The Theory of Sound*. New York: Dover.

Riazi, R., M. Farshchi, M. Shimura, M. Tanahashi, and T. Miyauchi. 2010. "An Experimental Study on Combustion Dynamics and NOx Emission of a Swirl Stabilized Combustor with Secondary Fuel Injection." *Journal of Thermal Science and Technology* 5 (2): 266–281. doi:10.1299/jtst.5.266.

Richards, G.A., D.L. Straub, and E.H. Robey. 2003. "Passive Control of Combustion Dynamics in Stationary Gas Turbines." *Journal of Propulsion and Power* 19 (5): 795–810. doi:10.2514/2.6195.

Romero-Jabalquinto, A., A. Velasco-Téllez, P. Zambrano-Robledo, and B. Bermúdez-Reyes. 2016. "Feasibility of Manufacturing Combustion Chambers for Aeronautical Use in Mexico." *Journal of Applied Research and Technology* 14 (3): 167–172. doi:10.1016/j. jart.2016.05.003.

Rosin, P., and E. Rammler. 1933. "Regularities in the Distribution of Cement Particles." *Journal of the Institute of Fuel* 7: 29–33.

Ryan, T. W., L. G. Dodge, and T. J. Callahan. 1984. "The Effects of Vegetable Oil Properties on Injection and Combustion in Two Different Diesel Engines." *Journal of the American Oil Chemists Society* 61 (10): 1610–1619. doi:10.1007/BF02541645.

Sallevelt, J. L. H. P., A. K. Pozarlik, M. Beran, L. U. Axelsson, and G. Brem. 2014. "Bioethanol Combustion in an Industrial Gas Turbine Combustor: Simulations and Experiments." *Journal of Engineering for Gas Turbines and Power* 136 (7): 071501. doi:10.1115/1.4026529.

Sano, Y., and R. B. Keey. 1982. "The Drying of a Spherical Particle Containing Colloidal Material into a Hollow Sphere." *Chemical Engineering Science* 37 (6): 881–889. doi:10.1016/0009-2509(82)80176-0.

Scarinci, T., Ch. Freeman, and I. Day. 2004. "Passive Control of Combustion Instability in a Low Emissions Aeroderivative Gas Turbine." *Proceedings of the ASME Turbo Expo 2004*, 1: 487–499 doi:10.1115/gt2004-53767.

Sirignano, W.A. 1999. *Fluid Dynamics and Transport of Droplets and Sprays.* Cambridge: Cambridge University Press. doi:10.1017/cbo9780511529566.

Sobulska, M. 2019. "Flame Spray Drying." PhD diss., Lodz University of Technology.

Sobulska, M., M. Piatkowski, and I. Zbicinski. 2017. "Flame Spray Drying: Droplet and Particle Flow Dynamics." *Drying Technology* 35 (8): 948–956. doi:10.108 0/07373937.2016.1233561.

Sobulska, M., I. Zbicinski, and M. Piatkowski. 2020. "Mechanism of Flame Spray Drying Process: Experimental and CFD Analysis." *Drying Technology* 38 (1–2): 80–92. doi:10. 1080/07373937.2019.1624566.

Strumillo, C., P. L. Jones, and R. Zylla. 2006. "Energy Aspects in Drying." In *Handbook of Industrial Drying*, edited by A. S. Mujumdar, 1075–1102. Boca Raton, FL: CRC Press.

Świątkowski, W. 1998. *Elementy Fizykochemii i Techniki Spalania Paliw : Dla Studentów Wydziału Inżynierii Chemicznej i Ochrony Środowiska.* Lodz: Technical University of Lodz.

Tran, T. T.H., M. Jaskulski, J. G. Avila-Acevedo, and E. Tsotsas. 2017. "Model Parameters for Single-Droplet Drying of Skim Milk and Its Constituents at Moderate and Elevated Temperatures." *Drying Technology* 35 (4): 444–464. doi:10.1080/07373937.2016.1182548.

Venkataraman, V., Y. Murai, M. Liverts, R. Örlü, J. H. M. Fransson, O. Stenlåås, and A. Cronhjort. 2020. "Resistance Wire Thermometers for Temperature Pulse Measurements on Internal Combustion Engines." In *SMSI 2020 Conference – Sensor and Measurement Science International*, 252–253. doi:10.5162/SMSI2020/P2.14.

Yang, Z. G., D. M. Paxton, K. S. Weil, J. W. Stevenson, and P. Singh. 2002. *Materials Properties Database for Selection of High-Temperature Alloys and Concepts of Alloy Design for SOFC Applications.* Richland, WA: Pacific Northwest National Laboratory.

Yarin, L. P., and G. Hetsroni. 2004. *Combustion of Two-Phase Reactive Media. Heat and Mass Transfer.* Berlin, Heidelberg: Springer. doi:10.1007/978-3-662-06299-9.

Yaws, C.L. 2009. "Yaws' Critical Property Data for Chemical Engineers and Chemists - Table 12. Vapor Pressure - Organic Compounds." https://app.knovel.com/web/view/itable/show.v/-rcid:kpYCPDCECD/cid:kt009ZN2S3/viewerType:eptble/root_slug:table-12-vapor-pressure---organic-compounds-log-p--a---bt--c/url_slug:table-12-vapor-pressure?

Yi, F., and R. L. Axelbaum. 2013. "Stability of Spray Combustion for Water/Alcohols Mixtures in Oxygen-Enriched Air." *Proceedings of the Combustion Institute* 34 (1): 1697–1704. doi:10.1016/j.proci.2012.05.088.

Yi, F., and R. L. Axelbaum. 2014. "Utilizing Preferential Vaporization to Enhance the Stability of Spray Combustion for High Water Content Fuels." *Combustion and Flame* 161 (8): 2008–2014. doi:10.1016/j.combustflame.2014.01.012.

Zbicinski, I. 2002. "Equipment, Technology, Perspectives and Modeling of Pulse Combustion Drying." *Chemical Engineering Journal* 86 (1–2): 33–46. doi:10.1016/S1385-8947 (01)00269-8.

Zbiciński, I., and M. Piątkowski. 2004. "Spray Drying Tower Experiments." *Drying Technology* 22 (6): 1325–1349. doi:10.1081/DRT-120038732.

Zonfrillo, G., M. Giovannetti, and I. Manetti. 2008. "Material Selection for High Temperature Applications." *Meccanica* 43: 125–131. doi:10.1007/s11012-008-9126-6.

2 Applications of Flame Spray Drying

2.1 APPLICATION OF FLAME SPRAY DRYING AND DRIED POWDER PROPERTIES

Sobulska (2019) successfully applied flame spray drying (FSD) process for drying of thermally resistant material, i.e., ceramic powder and thermally sensitive materials, such as maltodextrin and coffee. The author compared the results of FSD process with standard spray drying (SSD) in terms of final product properties, followed by analysis of effect of FSD process parameters on the key product properties such as PSD of dry powder, morphology, bulk and apparent density, moisture content, content of caramelization products (5-hydroxymethylfurfural, HMF), and color index.

2.2 CERAMIC POWDER

2.2.1 PARTICLE MORPHOLOGY

Ceramic material suspension (Ceramika Paradyz Inc., Opoczno, Poland), which is typically used for production of ceramic tiles, was chosen for comparison of novel FSD and SSD techniques. A comparison of the structure of ceramic material particles obtained after FSD and SSD is shown in Figure 2.1a and b, resp. For thermally resistant material, no agglomeration was observed for both FSD and SSD processes. Diameters of particles produced via FSD were ca. 80 μm, whereas diameters of particles produced by SSD were in the range of 60–90 μm. Generally, there are no significant differences in the PSD of ceramic particles obtained after FSD and conventional spray drying process.

- Effect of Fuel Concentration

    The morphology of ceramic powder obtained in FSD applying fuel concentrations of 25 and 50 wt.% is shown in Figure 2.2. Results of FSD show that the increase of ethanol concentration from 25 to 50 wt.% in the feed, which corresponds to decrease of air/fuel ratio from 112 to 48 during FSD, resulted in a slight reduction of particle diameters with an increase of the fraction of fine particles with diameters less than 50 μm. Intensive combustion of spray with high fuel concentration and consequently, lower amount of excess air result in partial damage of the particles.
- Effect of Fuel Type

    The morphology of ceramic material dried by FSD method for two fuel types: ethanol and sunflower oil is shown in the Figure 2.3. Application of

DOI: 10.1201/9781003100386-2

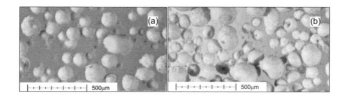

FIGURE 2.1   Ceramic particles (a) obtained by flame spray drying, (b) obtained by standard spray drying. (Piatkowski, Taradaichenko, and Zbicinski 2014 Permission Granted.)

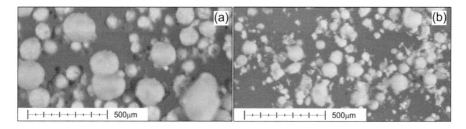

FIGURE 2.2   Ceramic particles after FSD, fuel content in the feed: (a) 25 wt.%, (b) 50 wt.%.

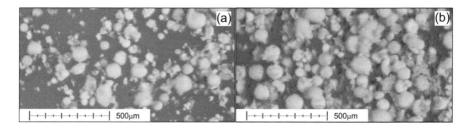

FIGURE 2.3   Ceramic particles after FSD, flammable component: (a) ethanol, (b) sunflower oil. (Piatkowski, Taradaichenko, and Zbicinski 2014 Permission Granted.)

ethanol as a flammable component resulted in production of ceramic particles in the range from ca. 80 to 10 µm. Particle diameters after FSD of ceramic material for both sunflower oil and ethanol are in the same range. For both fuel types, partially damaged particles with fractured pieces below ca. 10 µm are observed. In general, no significant differences concerning the product quality resulting from the fuel type are observed for ceramic powders.

2.3   MALTODEXTRIN

Sobulska (2019) carried out FSD test with maltodextrin, i.e., thermally sensitive and skin-forming material to analyze how different FSD process parameters affect material properties such as bulk density, PSD in the powder, particles agglomeration, puffing, and thermal degradation. Maltodextrin is a mixture of saccharides obtained from edible starch having a dextrose equivalency (DE) of less than 20, typically produced in the form of dried powder or purified aqueous solutions (Hobbs 2009). Maltodextrins may be produced by acid or by acid–enzyme conversion of starch. In FSD experimental tests, the authors used maltodextrin DE16 (Nowamyl S.A., Poland) characterized by the following parameters quoted by the producer: form – dried powder, color – white to slight cream, solubility – complete, mechanical impurities – absent, moisture content – 4.6%, pH – 6.2, ash content – 0.28% dry basis. Ash content, i.e., parameter commonly determined in the food products, shows number of nonorganic compounds in the sample.

2.3.1 PSD in the Dried Powder

- Effect of Atomization Pressure

  The effect of atomization pressure on cumulative PSD in the dried powder determined using image analysis technique is presented in Figure 2.4. Increase in dried particle AMD (arithmetic mean diameter) for lower atomization pressure is observed: for atomization pressure of 5.0 MPa, AMD was 109 μm and for atomization pressure of 0.7 MPa, AMD increased to 134 μm.

  Reduction of atomization pressure effects both product PSD and initial PSD in similar way: the lower the atomization pressure, the higher the initial particle diameters, which results in higher particle sizes in the product, (Figure 1.16).

- Effect of Solid Content in the Feed

  Results of the analysis of cumulative PSD of dried powder for different solid content are shown in Figure 2.5. Increase of solid content in the feed resulted in production of large particle diameters: dried powder AMD for solid content in the feed of 30, 40, and 50 wt.% was 49, 132, and 129 μm resp. Solid content in the feed affects initial PSD (Figure 1.22), for higher solid content in the feed, higher particle diameters in the product is obtained.

- Effect of Fuel Concentration

  The effect of fuel concentration in the feed on the PSD in the dried powder is shown in Figure 2.6. For lower fuel concentration, the increase of AMD of particles in the powder is observed; for fuel concentration of 35

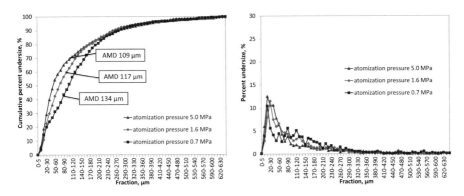

FIGURE 2.4 Effect of atomization pressure on the cumulative PSD of maltodextrin powder.

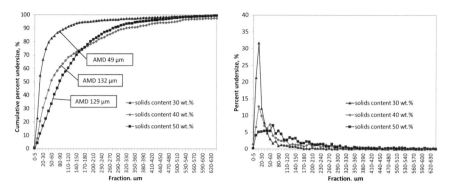

FIGURE 2.5   Effect of solid content in the feed on the cumulative PSD of maltodextrin powder.

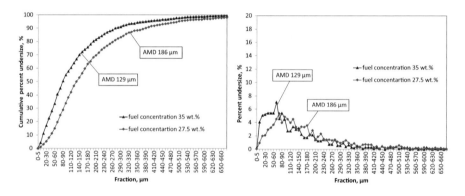

FIGURE 2.6   Effect of fuel concentration on the PSD of maltodextrin powder.

wt.%, AMD is 129 μm, whereas for fuel concentration of 27.5 wt.%, AMD increases to 186 μm. As shown earlier (Figure 1.23), decrease of fuel content in the feed resulted in bigger initial particle diameters and finally in higher AMD of particles in the dried product.

• Effect of Feed Temperature

Figure 2.7 shows the effect of feed temperature on the PSD in the dried powder. Increase of feed temperature from 30°C to 70°C slightly increases AMD in dried powders from 49 to 61 μm. Higher feed temperature affects initial PSD (Figure 1.24), higher initial AMD of the particles results in bigger particles in the product.

Relationship between dried particle AMD obtained in FSD tests and average particle residence time in flame was found. AMD of dried powder for different particle residence time in the flame is shown in Table 2.1. Particle residence time in the flame depends on the following parameters: initial particle velocity and length of the flame controlled by atomization parameters and composition of dried solution. Increase of dried particles

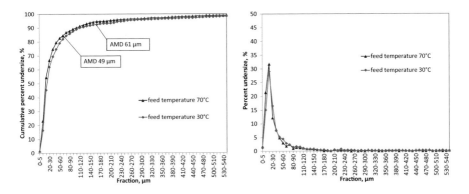

FIGURE 2.7   PSD in dried powder for different feed temperatures of 30°C and 70°C.

TABLE 2.1
Powder AMD and Particle Residence Time in the Flame

Particles AMD (µm)	Particles Residence Time in Flame (s)
49	0.02
61	0.02
83	0.04
132	0.03
135	0.05
109	0.09
117	0.06
134	0.08
92	0.06
129	0.22
186	0.16

Source: Based on Sobulska, Zbicinski, and Piatkowski (2020).

AMD with increase of residence time in flame due to particle puffing in the high-temperature flame is observed.

2.3.2   PARTICLES MORPHOLOGY

Figure 2.8 shows morphology of dried maltodextrin particles obtained in (a) FSD and (b) SSD process. In the FSD process, agglomerates are produced due to particles flow in the recirculation zones on the edges of the flame. The formation of agglomerated particles is not typical for the co-current spray drying system due to parallel flows of

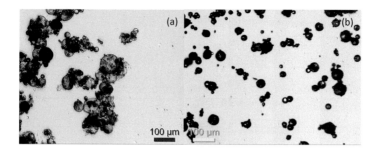

FIGURE 2.8 Morphology of dried particles obtained by (a) FSD process; (b) standard spray drying process.

FIGURE 2.9 Maltodextrin particles after (a) FSD and (b) standard spray drying, SEM (Piatkowski, Taradaichenko, and Zbicinski 2015 Permission Granted.)

drying air and dispersed particles (Zbiciński and Piątkowski 2004). Agglomeration improves instant product properties such as porosity, wettability, solubility, and flowability of the dried powder (Dhanalakshmi, Ghosal, and Bhattacharya 2011).

Scanning electron microscope (SEM) images of maltodextrin particles obtained in the FSD were compared with powders produced in the SSD. Figure 2.9 shows a significant number of fractured maltodextrin particles in the SSD process due to high outlet air temperatures (132°C); similar effects were reported by Kwapińska and Zbiciński (2005). In powder produced in the FSD process, the content of fractured particles was lower than in SSD due to short drying time.

- Effect of Atomization Pressure

    Figure 2.10 shows microscopic images of maltodextrin particles obtained for different atomization pressure. The images show that decrease of the atomization pressure results in increase of particle diameters due to puffing caused by longer particle residence time in the high-temperature area of the flame. For example, for a test with atomization pressure of 5.3 MPa, an average particle residence time in flame is 0.03 s, for a test with atomization pressure of 1.6 MPa – 0.06 s, and for a test with atomization pressure of 0.7 MPa – 0.08 s. For tests carried out at atomization pressures of 0.7 and 0.3 MPa, the large number of fractured particles is observed due to particle overheating at high flame temperature.

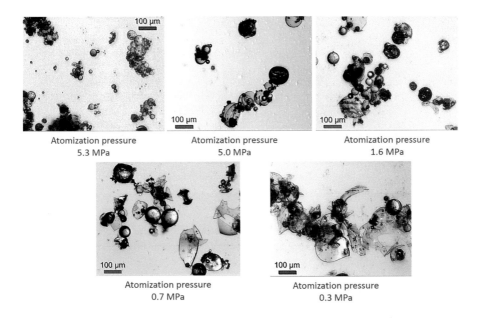

Atomization pressure
5.3 MPa

Atomization pressure
5.0 MPa

Atomization pressure
1.6 MPa

Atomization pressure
0.7 MPa

Atomization pressure
0.3 MPa

FIGURE 2.10 Morphology of maltodextrin powder for different atomization pressure.

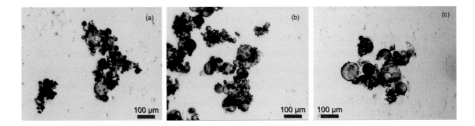

FIGURE 2.11 Morphology of maltodextrin powder obtained in tests with different solid content in the fuel-free solution: (a) 30 wt.%; (b) 40 wt.%; (c) 50 wt.%.

- Effect of Solid Content

    The effect of solid content in the feed on maltodextrin particle morphology is illustrated in Figure 2.11, where results for 30, 40, and 50 wt.% of solid in the feed are presented. In tests with 30 wt.% solids (fine spray, short flame), a big amount of small non-puffed particles is observed along with large agglomerates consisting of small primary particles (Figure 2.11a). In the aforementioned tests, the temperature in the drying tower does not exceed 160°C, and particle residence time in flame is short, 0.02 s; therefore, particle puffing is not observed. In other tests, where average temperatures in the flame zone are higher and particle residence time in flame is longer, 0.03 s for solid content of 40 wt.% and 0.22 s for solid content of 50 wt.%, particle inflation occurs. Comparing data of temperature distributions

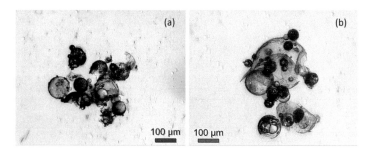

FIGURE 2.12 Morphology of maltodextrin powder obtained in tests with different fuel (ethanol) concentration: (a) 35 wt.%; (b) 27.5 wt.%.

in the tower (Figure 1.41) and particle morphology, the influence of high temperatures in the combustion zone, the formation of large particles due to puffing are observed.

- Effect of Fuel Concentration

Figure 2.12 presents particle agglomerates with inflated and fractured particles of maltodextrin powder obtained for different fuel concentrations in the feed. Large particles with diameters > 150 μm (AMD 186 μm) are found for lower fuel concentration of 27.5 wt.% in relation to 35 wt.% (AMD 129 μm) for the similar flame length. For lower fuel concentration, an increase of particle diameters is caused by higher viscosity of the solution, which produces higher initial AMD of particles.

Analyzing SEM photos presented in Figure 2.13a–c, strong effect of fuel concentration in the feed on PSD can be observed. Figure 2.13a shows that for 35 wt.% of fuel, large particles with a rough external surface are formed. Increase of fuel concentration in the feed to 40–45 wt.% (Figure 2.13b and c) produces spherical particles with increased fraction of fines. AMD of the dried powder decreased from 69 μm (fuel concentration 35 wt.%) to 43 and 32 μm with increase of fuel concentration to 40 and 45 wt.%, resp.

As fuel concentration in the feed does not affect flame length and particle residence time in flame, differences in particle morphology are caused by lower feed viscosity and fine atomization (lower initial AMD) for higher fuel concentration.
- Effect of Air Flow Rate in the Tower

Figure 2.14 shows particle morphology for tests with different air flow rates of 340 and 480 Nm³/h in the tower (Sobulska 2019). Particles dried in both tests are puffed and partially fractured. Comparison of the microscopic images shows that for higher air flow rate (480 Nm³/h), higher number of caramelized particles can be found for similar flame length. For higher air flow rate (480 Nm³/h), particle velocities in the tower are higher, which results in slight decrease of particle residence time in flame, but because of higher flame temperature in the vicinity of the nozzle (Figure 1.43), the particles are overheated. Particle AMD increases from 92 to 134 μm with

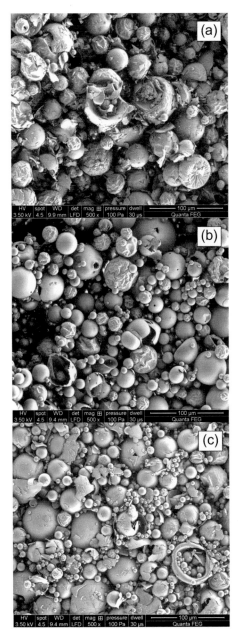

FIGURE 2.13 Maltodextrin particles after FSD, SEM, fuel content in the feed: (a) 35 wt.%, (b) 40 wt.%, (c) 45 wt.% (Piatkowski, Taradaichenko, and Zbicinski 2015 Permission Granted.)

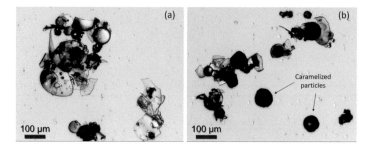

FIGURE 2.14 Microscopic images of maltodextrin particles for different air flow rate in the tower: (a) 340 Nm³/h; (b) 480 Nm³/h.

decrease of air flow rate in the tower due to longer residence time in the flame (air flow rate of 340 Nm³/h – 0.06 s, air flow rate – 480 Nm³/h – 0.08 s) causing particle puffing.

2.3.3 BULK AND APPARENT DENSITY, MOISTURE AND FUEL CONTENT, COLOR INDEX, AND HMF CONTENT

Extensive experiments were carried out in Sobulska (2019), which enabled to evaluate the effect of FSD process parameters on the dried powder properties: bulk and apparent density, final moisture content of the powder, and content of remaining fuel (ethanol). To evaluate thermal degradation of the product, parameters such as color index and 5-hydroxymethylfurfural (HMF) content can be used. HMF is the main product of caramelization reaction taking place in the carbohydrates subjected to high temperatures (Kroh 1994). Content of HMF in the dried powder can be determined applying colorimetric Winkler method (Martysiak-Żurowska and Borowicz 2009; Winkler 1955).

Table 2.2 displays bulk and apparent density, moisture content, color index, and HMF content of the maltodextrin powders obtained after FSD and SSD.

TABLE 2.2

Comparison of Dried Powder Properties for Flame Spray Drying (FSD) and Standard Spray Drying (SSD)

Test	Fuel Concentration in the Feed (wt.%)	Solid Content in the Fuel-Free Solution (wt.%)	Bulk Density (kg/m³)	Apparent Density (kg/m³)	Moisture Content in Powder (wt.%)	Color Index	HMF Content (mg/kg)
SSD	0	30	567	1,410	7.58	6.7	42.9
FSD	35	30	228	1,021	5.86	10.6	116.1
SSD	0	40	588	1,358	5.99	7.2	7.5
FSD	35	40	153	937	3.50	18.7	116.2

FIGURE 2.15 Color index for different maltodextrin samples after dissolution (concentration of maltodextrin in each sample is 20 wt.%).

Comparison of the results shows that the application of combustion process as an energy source for drying affects dried powder properties. Powder produced after FSD process has bulk density of 228 and 153 kg/m³ and apparent density – 1,021 and 937 kg/m³, whereas in SSD process, bulk density is in a range from 567 to 588 kg/m³ and apparent density from 1,358 to 1,410 kg/m³. The bulk and apparent densities of powder produced by FSD are lower compared with SSD due to particle puffing and agglomeration. Low bulk density of the powder can be a disadvantage due to increase in transport and storage costs; however, in certain cases, low bulk and apparent densities of the powder are required e.g., low bulk density and high porosity are preferable for detergent powder (Francia et al. 2016).

Color index of maltodextrin powder solutions produced by FSD and SSD process can be determined using colorimetric method described by Hobbs (2009). Color indexes for maltodextrin solutions obtained in different SSD and FSD processes are shown in Figure 2.15.

For SSD process, color index of the samples is equal to 6.7 and 7.2 (Figure 2.15). For FSD test with solid content of 30 wt.%, slight increase of powder color index to 10.6 is observed, but in test with solid content of 40 wt.%, partial caramelization of the sample was detected as color index raised to 18.7. Analysis of HMF content in dried powder shows that after SSD, content of caramelized product in samples is in the range from 7.5 mg/kg to 42.9 mg/kg, whereas after FSD, HMF increases to 116 mg/kg in corresponding tests. Higher HMF content in powder for FSD results from caramelization processes due to particles overheating in the flame.

- Effect of Atomization Pressure

    Table 2.3 shows the effect of atomization pressure on dried powder properties. Bulk density of the powder varies from 99 to 198 kg/m³ and apparent density from 898 to 1,158 kg/m³. Bulk and apparent densities of the powder slightly increase (high number of fractured particles (Figure 2.10)) with decrease of atomization pressure due to longer particle residence time in the high-temperature environment. Decrease of atomization pressure from 5.3 MPa to 1.6 and then to 0.7 MPa results in increase of color index from 18.7 to 33.8 and 41.5 due to increase of particle residence time in flame from 0.03 to 0.06 and 0.08 s. Increase of HMF content from 116.2 to 171.3 mg/kg for atomization pressure of 5.3 and 1.6 MPa, resp., is also observed due to increase of particle residence time in flame and increase of flame temperature for lower atomization pressure. Decrease of atomization pressure results in lower moisture content in the dried powder due to higher

TABLE 2.3

Effect of Atomization Pressure on the Dried Powder Properties (FSD)

Atomization Pressure (MPa)	Nozzle Type/ Feed Rate	Bulk Density (kg/m³)	Apparent Density (kg/m³)	Moisture Content (wt.%)	Ethanol Content (wt.%)	Color Index	HMF Content (mg/kg)
5.3	Fine 0.6 10 kg/h	153	937	3.50	1.47	18.7	116.2
5.0	Fine 0.4 7 kg/h	164	898	4.69	1.83	20.1	133.4
1.6	Fine 0.6 5.5 kg/h	176	1,026	3.09	1.34	33.8	171.3
0.7	ConeJet 0.6 5.5 kg/h	198	1,158	2.21	2.12	41.5	97.1
0.3	Fine 1.0 10 kg/h	99	1,062	2.16	0.90	70.4	220.1

TABLE 2.4

Effect of Solid Content in the Feed on the Dried Powder Properties

Fuel Concentration in the Feed (wt.%)	Solid Content in the Fuel-Free Solution (wt.%)	Bulk Density (kg/m³)	Apparent Density (kg/m³)	Moisture Content in Powder (wt.%)	Ethanol Content in Powder (wt.%)	Color Index	HMF Content (mg/kg)
35	30	228	1,021	5.86	0.87	10.5	116.1
35	40	153	937	3.50	1.47	18.7	116.2
35	50	103	1,017	2.57	1.97	84.9	228.3

temperatures in the combustion zone and more intensive drying. Final ethanol content in the dried particles varies from 0.87 to 2.12 wt.%.

- Effect of Solid Content

    The effect of solid content in the feed on the dried powder properties such as bulk and apparent density, moisture content, ethanol content, color index, and HMF content is presented in the Table 2.4. Increase of solid content in the feed results in lower bulk density of powder due to longer residence time of droplets in the high-temperature zone, which promotes puffing of particles and lowers product bulk density. Analysis of color index and HMF content for tests with different solid content of 30, 40, and 50 wt.% shows that higher solid content in the feed increases color index and HMF content. For solid content of 50 wt.%, the longest residence time in flame was about 0.22 s, which elevated the color index to 84.9. Reduction of solid content to 40 wt.% resulted in shorter flame length, for which particle residence time in the flame of 0.03 s decreased color index to 18.7. Similar observations are valid for the effect of solid content in the feed on HMF content in the powder.

Increase of ethanol content in the powder for higher solid content observed in the maltodextrin powder agrees with the study of Furuta, Okazaki, and Toei (1983), who reported higher ethanol retention in maltodextrin particles for higher initial solid content and for higher drying air temperatures. The authors explained that for higher initial solid content and higher drying air temperatures, the critical moisture content of the droplets is achieved faster and ethanol is encapsulated in dried crust of solid material. Relationship between solid content in the feed and final moisture content of the powder can also be observed: powder moisture content decreases with increase of solid content in the feed.

- Effect of Fuel Concentration

    Table 2.5 shows final product properties for different fuel concentrations in the feed. The bulk density of powder ranged from 100 to 228 kg/m³ and apparent density varied from 793 to 1,021 kg/m³. There is no clear relationship between fuel concentration in the feed and bulk and apparent density of the dried powder. Decrease of moisture content in the dried powder is observed for higher fuel content in the feed. Increase of color index and HMF content in the powder indicates more intensive particle degradation caused by lower air/fuel ratio. Ethanol content in the dried powder varied from 0.77 to 2.28 wt.%, no relationship between fuel concentration and ethanol content is found.

- Effect of Air Flow Rate

    Table 2.6 shows dried product properties for two air flow rates. The enhanced air supply in the combustion zone and higher flame temperatures for air flow rate 480 Nm³/h resulted in higher powder bulk and apparent densities due to increased number of fractured and caramelized particles (Figure 2.14). For tests with different air flow rate in the tower flame length was similar; however, increase of air flow rate from 380 to 480 Nm³/h affected particle velocities, which resulted in decrease of

---

TABLE 2.5

Effect of Fuel Concentration in the Feed on the Dried Powder Properties

Fuel Concentration in the Feed (wt.%)	Solid Content in the Fuel-Free Solution (wt.%)	Bulk Density (kg/m³)	Apparent Density (kg/m³)	Moisture Content in Powder (wt.%)	Ethanol Content in Powder (wt.%)	Color Index	HMF Content (mg/kg)
35	30	228	1,021	5.86	0.87	10.5	116.1
45	30	145	793	3.41	1.14	29.5	130.5
35	40	153	937	3.50	1.47	18.7	116.2
45	40	126	849	1.46	0.77	77.4	370.5
27.5	50	100	920	2.68	2.28	73.5	56.5
35	50	103	1,017	2.57	1.97	84.9	228.3

TABLE 2.6

Effect of Air Flow Rate in the Tower on the Dried Powder Properties

Air Flow Rate (Nm³/h)	Nozzle Type	Feed Rate (kg/h)	Bulk Density (kg/m³)	Apparent Density (kg/m³)	Moisture Content (wt.%)	Ethanol Content (wt.%)	Color Index
340	ConeJet 0.6	5.5	198	1158	2.21	2.12	41.5
480	ConeJet 0.6	5.5	249	1234	2.55	2.32	37.3

TABLE 2.7

Effect of Feed Temperature on the Dried Powder Properties

Feed Temperature (°C)	Atomization Pressure (MPa)	Bulk Density (kg/m³)	Apparent Density (kg/m³)	Moisture Content (wt.%)	Ethanol Content (wt.%)	Color Index
70	6.2	228	1,021	5.86	0.87	10.5
30	5.5	202	908	4.75	1.21	10.6

particle residence time in flame from 0.08 to 0.06 s and decrease of color index from 41.5 to 37.3.

- Effect of Feed Temperature

Table 2.7 shows dried powder properties for two initial temperatures of the feed. Decrease of feed temperature from 70°C to 30°C produced slightly lower bulk (228 and 202 kg/m³ resp.) and apparent densities (1,021 and 908 kg/h) caused by change of initial atomization parameters (higher feed viscosity). Particles produced in the tests with feed temperature 70°C have lower diameters and are well packed in the material bed, decreasing the number of voids between particles, which increases bulk density of the product.

Feed temperature in FSD does not have significant influence on the dried powder moisture content, ethanol content, and color index as particle residence time in flame (0.02 s) is the same.

2.4 FSD OF COFFEE

Another thermally sensitive product that was dried applying FSD process was soluble coffee (Lodz University of Technology, 2015, unpublished data). Feed solution for drying tests was prepared by dissolving commercially available granulated coffee (Nestle, Poland) in water achieving solid content in the solution of 50 wt.%. After complete dissolution of coffee flammable component, i.e., ethanol was added to the solution to obtain fuel concentration in the feed of 45 wt.%.

Flame spray dried coffee powder was collected and analyzed to determine particle size distribution, morphology, bulk and apparent density, moisture, and ethanol content.

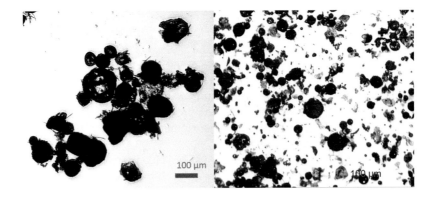

FIGURE 2.16   Microscopic image of soluble coffee powder obtained by FSD.

Figure 2.16 shows morphology of coffee particles obtained via FSD process. The structure of observed particles is highly differentiated: some of particles are single non-agglomerated, whereas particle agglomerates are also present, some of the particles comprise internal voids, and there is a high amount of fractured particles. The Sauter mean diameter (SMD) of dry coffee particles determined by laser diffraction analyzer (Spraytec, Malvern Instruments, UK) is 128 μm, whereas values of D_{v10}, D_{v50}, and D_{v90} were 67, 225, and 590 μm, respectively. The bulk and apparent densities of FSD coffee powder are slightly higher compared with maltodextrin powders: bulk density of coffee – 297 kg/m^3 and apparent density – 1,378 kg/m^3. Moisture (8.71% wet basis) and ethanol content (5.25%) of coffee powder are significantly higher compared with maltodextrin powder obtained by FSD process (Tables 2.2–2.7).

2.5   SUMMARY AND STRATEGIES TO CONTROL THE PRODUCT QUALITY

The properties of dried powder may be controlled in FSD process.

Decrease of atomization pressure produces longer flame, which increases particle residence time in flame causing puffing, higher AMD and SMD, reduction of moisture content in the product, increase of color index and content of caramelized products in the powder.

Higher solid content in the feed increases particle puffing leading to bigger AMD and SMD, lower bulk density, increased ethanol content, higher color index and HMF content due to longer particle residence time in flame.

Lower fuel content in the feed and low feed temperature result in increase of initial viscosity of the solution producing higher initial and final particle AMD and SMD. Increase of fuel content in the feed results in higher color index and HMF content in the powder due to intensive particles degradation in high flame temperature caused by lower air/fuel ratio.

Increase of air flow rate in the tower results in increase of flame temperatures in the core of the spray near the nozzle. Therefore, increased number of fractured particles is found for higher air flow rate in the tower. Higher air flow rate in the tower reduces particle residence time in flame due to higher particle velocities and results in decrease of color index of the product.

The strategy of how to control FSD process and final product properties on the basis of the interactions between FSD process parameters (initial droplets AMD and SMD, spraying angle, flame temperature, AMD and SMD along the dryer height) and product properties (AMD in the product, bulk and apparent density, moisture content, ethanol content, color index, and HMF content) is presented in Table 2.8.

In the Table 2.8, the change of process or product characteristic is shown with the arrows "↓" – decrease, "↑" – increase or "–" – lack of observed relationship as a function of: atomization pressure, feed rate, solid content in the feed, fuel concentration in the feed, air flow rate in the tower, feed temperature.

TABLE 2.8
Strategies to Control the Product Quality

	Process Parameters (↑ – Increase)					
Characteristic of Spray or Product	Atomization Pressure (MPa ↑)	Feed Rate (kg/h ↑)	Solid Content in the Feed (wt.% ↑)	Fuel Concentration in the Feed (wt.% ↑)	Air Flow Rate in the Tower (kg/h ↑)	Feed Temperature (°C ↑)
Initial AMD, SMD (μm)	↓	↓	↑	↓	–	↓
Spraying angle (°)	↑	↑	↓	↑	–	↑
Flame temperature (°C)	↓	↓	↑	↑	↑↓	–
AMD, SMD along the dryer height (μm)	↓	↓	↑	↓	–	↓
AMD in the product (μm)	↓	↓	↑	↓	–	↓
Bulk density (kg/m³)	↓	↓	↓	–	↑	↑
Apparent density (kg/m³)	↓	↓	–	–	↑	↑
Moisture content (mass%)	↑	↑	↓	↓	–	–
Ethanol content (mass%)	–	–	↑	–	–	–
Color index	↓	↓	↑	↑	↓	–
HMF content (mg/kg)	↓	↓	↑	↑	–	–

The process–product interactions were elaborated based on experimental data on maltodextrin drying with addition of ethanol as a flammable component of FSD process.

ACRONYMS

AMD – Arithmetic Mean Diameter
FSD – Flame Spray Drying
HMF – Hydroxymethylfurfural
PSD – Particle Size Distribution
SEM – Scanning Electron Microscope
SMD – Sauter Mean Diameter
SSD – Standard Spray Drying

REFERENCES

Dhanalakshmi, K., S. Ghosal, and S. Bhattacharya. 2011. "Agglomeration of Food Powder and Applications." *Critical Reviews in Food Science and Nutrition* 51 (5): 432–441. doi:10.1080/10408391003646270.

Francia, V., L. Martín, A. E. Bayly, and M. J. H. Simmons. 2016. "Agglomeration in Counter-Current Spray Drying Towers. Part B: Interaction between Multiple Spraying Levels." *Powder Technology* 301: 1344–1358. doi:10.1016/j.powtec.2016.05.010.

Furuta, T., M. Okazaki, and R. Toei. 1983. "Effect of Drying on Retention of Ethanol in Maltodextrin Solution During Drying of a Single Droplet." *Drying Technology* 2 (3): 311–327. doi:10.1080/07373938308959834.

Hobbs, M. 2009. "Sweeteners from Starch: Production, Properties and Uses." In *Starch: Chemistry and Technology*, edited by J. BeMiller and R. Whistler, 797–832. London: Academic Press.

Kroh, L.W. 1994. "Caramelisation in Food and Beverages." *Food Chemistry* 51 (4): 373–379. doi:10.1016/0308-8146(94)90188-0.

Kwapińska, M., and I. Zbiciński. 2005. "Prediction of Final Product Properties after Cocurrent Spray Drying." *Drying Technology* 23 (8): 1653–1665. doi:10.1081/DRT-200065075.

Martysiak-Żurowska, D., and A. Borowicz. 2009. "A Comparison of Spectrophotometric Winkler Method and HPLC Technique for Determination of 5-Hydroxymethylfurfural in Natural Honey." *Chemia Analityczna* 54 (5): 939–947.

Piatkowski, M., M. Taradaichenko, and I. Zbicinski. 2014. "Flame Spray Drying." *Drying Technology* 32 (11): 1343–1351. doi:10.1080/07373937.2014.903413.

Piatkowski, M., M. Taradaichenko, and I. Zbicinski. 2015. "Energy Consumption and Product Quality Interactions in Flame Spray Drying." *Drying Technology* 33 (9): 1022–1028. doi:10.1080/07373937.2014.924137.

Sobulska, M. 2019. "Flame Spray Drying." PhD diss., Lodz University of Technology.

Sobulska, M., I. Zbicinski, and M. Piatkowski. 2020. "Mechanism of Flame Spray Drying Process: Experimental and CFD Analysis." *Drying Technology* 38 (1–2): 80–92. doi:10.1080/07373937.2019.1624566.

Winkler, O. 1955. "Beitrag Zum Nachweis Und Zur Bestimmung von Oxymethylfurfurol in Honig Und Kunsthonig." *Zeitschrift Für Lebensmittel-Untersuchung Und -Forschung* 102 (3): 161–167. doi:10.1007/bf01683776.

Zbiciński, I., and M. Piątkowski. 2004. "Spray Drying Tower Experiments." *Drying Technology* 22 (6): 1325–1349. doi:10.1081/DRT-120038732.

3 Flame in Drying and Particle Synthesis Techniques

3.1 FLAME DRYING OF TEXTILE

The first method utilizing combustion of a flammable liquid as an energy source for a drying process was elaborated in the 1972 to dry textiles, also known in the literature as the Remaflam® process (Birke, Eltz, and Schön 1972; Eltz, Petersohn, and Schön 1981).

In this process, the fabric is soaked in the aqueous solution of an organic solvent, such as methanol, ethanol, or isopropanol, ignited, and then dried in the flame. The flammable components were selected based on the safety limits values of staff exposure to chemicals and flash points. For example, the maximum 8 h average workplace exposure to ethanol is 1,000 ppm TWA (time weighted average), to methanol is 200 ppm TWA, and to isopropanol is 400 ppm TWA, which was acceptable for flame drying process, whereas dioxane with lower exposure limit of 50 ppm has been considered as unsuitable due to safety issues (Eltz, Petersohn, and Schön 1981). Analysis of flash points (FPs) for common organic solvents showed that ethanol (FP = 12°C), methanol (FP = 11°C), and isopropanol (FP = 13°C) could be applied as a fuels in Remaflam® process, whereas acetone has extremely low FP of −20°C and could not be applied as a fuel due to safety reasons (Eltz, Petersohn, and Schön 1981).

The Remaflam® process includes the following operations: initially, the water solution containing dye is mixed with about 35 vol.% of flammable component (methanol), then the fabrics is soaked in the water/dye/fuel mixture and supplied to the drying chamber. Wet fabric is supplied from the bottom of the drier, where IR heaters were installed to heat up the water/dye/methanol mixture above the FP (35°C) to initiate the combustion. Cold air needed for combustion process is supplied at the bottom and the top of the drier by ventilators and fed to the combustion chamber. The authors reported that the temperature of the fabrics during simultaneous combustion and drying does not exceed 70°C due to heat absorption during the moisture evaporation (Eltz, Petersohn, and Schön 1981).

Figure 3.1 shows the scheme of flame dryer of textile that was patented in Germany in 1972 (Birke, Eltz, and Schön 1972). The fabric containing flammable component and dye solution is supplied to the drying chamber, where igniters located on both sides of the fabric initiate the combustion of flammable component. The cool air required for combustion process is supplied through the air inlets located at the bottom of the dryer, and flue gases are removed from the top of the dryer. Due to safety reasons, the fabrics inlet and outlet to the combustion chamber has been sealed by

DOI: 10.1201/9781003100386-3

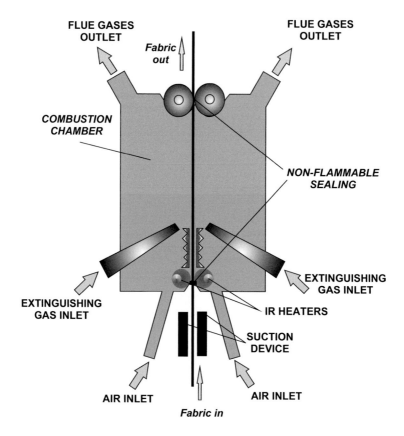

FIGURE 3.1   Schematic of flame dryer of textile. (Based on Birke, Eltz, and Schön 1972.)

entry slots produced from nonflammable materials. To avoid the uncontrollable ignition of fuel vapors on the fabric surface, which may occur before fabric enters the
combustion chamber, the suction device may be installed on both sides of the fabric
to remove the flammable vapors before the combustion chamber inlet. Two nozzles
supplying nonflammable gas may be applied to extinguish the flame inside the combustion chamber in the case of emergency.

Later, in 1981, Eltz, Petersohn, and Schön presented a Remaflam® dryer, which has
been constructed and operated in the technical center of the ATA dyes of Hoechst AG
in Germany. Figure 3.2 shows the scheme of the Remaflam® dryer, the flow direction
of the hot and cold gases, and location of process control elements. The thermocouple
located in the combustion chamber controls the rotation speed of the outlet gas fan
setting the required air flow rate in the chamber. The temperature of the fabric during
combustion process may be kept below 100°C only if some moisture is remaining in
the fabric; therefore, to avoid fabrics overheating, the moisture measuring device was
installed at the fabrics outlet, which controls the speed throughput of the fabric in the
drying chamber. Additionally, UV cell has been installed in the combustion chamber
to detect the flame; if for some reason the flame is not detected, the control system

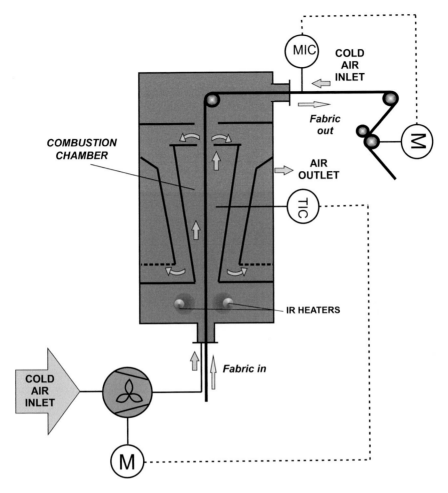

FIGURE 3.2   Flame drying of textile – Remaflam® process. (Based on Eltz, Petersohn, and Schön 1981.)

will shut down the drive of fabric supplying elements. The air supplying fans ensure continuous ventilation of the combustion chamber, preventing the accumulation of the explosive fuel vapors in the combustion chamber.

Table 3.1 shows the data on process parameters recorded during drying of polyester and cotton blend fabric of weight 320 g/m² with methanol applied as a fuel at concentration of 34 vol.% (Eltz, Petersohn, and Schön 1981). The fabric has been dried with the optimal throughput speed of 40 m/min to keep the final moisture content of the fabric at ca. 4%. The air flow rate in the chamber was 6000 m³/h and the temperature in the combustion chamber was 600°C, whereas the temperature of the flue gases was about 800°C. The flame drying process has also been applied to dry textiles such as cotton, polyester, and blends. The authors mentioned that textile ability to absorb the feed solution containing fuel is also an important parameter determining the applicability of Remaflam® process. Additionally, Table 3.1 shows throughput

TABLE 3.1

Process Parameters Data Recorded During Remaflam® Drying of Different Types of Textiles

Parameters	Polyester/ Cotton Blend	Polyester/Cotton Blend (Ratio 67/33)	Polyester/Cotton Blend (Ratio 50/50)	Cotton
Methanol content	34 vol.%	34 vol.%	34 vol.%	34 vol.%
Solution absorption during fabric soaking (%)	nd	42	50	70
Linear density (g/m²)	320	230	110	230
Textile throughput speed (m/min)	ca. 40	45	60	23
Temperature in the combustion chamber (°C)	600	600	600	600

Source: Based on Eltz, Petersohn, and Schön (1981).

speed and feed solution absorption achieved during flame drying of different types of fabrics (Eltz, Petersohn, and Schön 1981).

The authors (Eltz, Petersohn, and Schön 1981) proved that thermal efficiency of the Remaflam® process was around 72%, whereas in the conventional textile drying process, where drying air was heated up by oil combustion, efficiency was only 50%. Other advantages of Remaflam® dryer are possibilities for simultaneous dyeing and drying as well as small area required for installation (Hoverath 1981).

More recently, in 2009, Robin and Lenoir had patented a flame dryer, which could be applied for dewatering of thin sheet materials, for example, paper (Robin and Lenoir 2009). In the invented dryer, the wet paper is passed through the flame generated from specially designed gas burners known as Furinit® (Bekaert Combustion Technology B.V.). Application of flame drying at high temperatures in the range from 600 to 1,000°C allows to reduce the drying time and increase the throughput speed of the paper to the range of 50–2,000 m/min, which allows for the dryer capacity increase. The gas burners are located at the distance of 0.5–10 cm from the passing wet paper in the sequence with hot gas blowing nozzles. The nozzles are installed to blow the flue gases generated during combustion process on the paper and reuse the waste energy to enhance the efficiency of paper drying process. The process is continuously controlled by the paper moisture content at the dryer outlet via increase or reduction of the paper throughput speed or flame temperature.

Nowadays, flame treatment of the textiles has found an industrial application for singeing process, i.e., removal of short fibers from the textile surface to enhance further processing operations such as dyeing, impregnation, etc. Several manufacturers of textile processing equipment offer flame singeing machines, where undesired fibers are burnt out from the textile surface by means of gas burners. For example, Cibitex Srl (Solbiate Olona, Italy) offers flame singeing line, which could process woven or knitted fabrics. Electronics for Imaging, Inc. (Fremont, CA, United States) offers flame singeing mashing designated for processing of heat sensitive polyester fabrics as well as natural cotton fabrics at the maximum textile speed of 120 m/min. Other main producers of flame singeing machines are Osthoff-Senge GmbH & Co. KG (Germany), Swastic (India), Menzel (USA), etc.

3.2   PULSE COMBUSTION DRYING

3.2.1   THE PRINCIPLE OF PULSE COMBUSTION

Pulse combustion (PC) originally was elaborated for propulsion and combustion applications about 90 years ago and has currently gained a high research interest for application in the drying processes (Mujumdar 2004; Zhonghua and Mujumdar 2006). PC burners, contrary to conventional continuous combustion burners, operate in the intermittent or periodic mode, where fuel burns in cycles generating pressure, velocity, and temperature oscillations (Zbicinski, Benali, and Kudra 2002). The cycle of pulse combustor consists of the following steps: fuel and air enter the combustion chamber and burns generating the pressure increase forcing the flue gases to escape the combustion chamber through tailpipe. As the result of the outflow, the pressure in the combustion chamber decreases and new portions of fuel and air are sucked into the combustion chamber, and the cycle is repeated (Zbicinski, Kudra, and Liu 2014).

Depending on the design of fuel and air inlet section, pulse combustors may be divided into two types: valved pulse combustors with mechanical membrane or rotary valves and valveless pulse combustors with inertial gas valve (Zbicinski 2002). Based on the type of the acoustic system applied, pulse combustors are classified into three groups: Schmidt (quarter valve) PC, Helmholtz PC, and Rijke-tube PC (Kudra 2008). Commonly, pulse combustors operate at the following process parameters: input power from 20 to 1,000 kW, frequencies from 20 to 500 Hz, pressure fluctuation of ± 10 kPa, and velocity fluctuations of $\pm$ 100 m/s in the combustion chamber.

PC technology may be applied in the drying process such as spray drying, fluidized bed drying, and flash drying (Zhonghua and Mujumdar 2006). In the PC spray dryer, the outlet of the PC tailpipe is connected to the drying chamber as shown in Figure 3.3.

3.2.2   ADVANTAGES AND DISADVANTAGES OF PULSE COMBUSTION DRYING

Application of pulse combustion in drying processes offers the following benefits over the conventional drying systems:

- Increase of the heat transfer coefficient by a factor of 2–5 due to high-intensity velocity fluctuations (Zbicinski, Benali, and Kudra 2002; Zbicinski 2002);
- Increase of the heat transfer area due to enhanced atomization of liquid feedstock promoted by intensive acoustic waves generated in the pulse combustor (Zbicinski 2002). In PC spray dryer, it is possible to obtain smaller droplet diameters with narrow particle size distribution without application

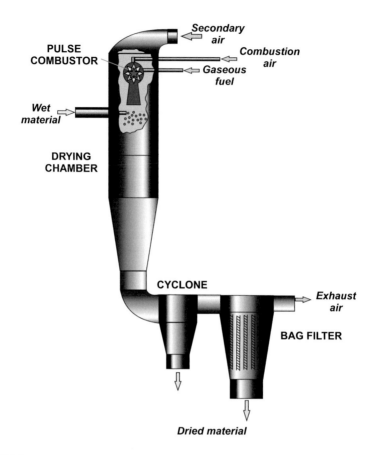

FIGURE 3.3 Schematic of the pulse combustion spray dryer. (Based on Wawrzyniak, Zbicinski, and Sobulska 2017.)

of atomizing device such as nozzles or rotation disks (Wang, Cui, and Sunada 2007). The PC dryer can atomize the slurries with viscosities up to 16,000 cP and solutions with viscosities up to 300 cP, which covers the wide range of materials (Kudra 2008).

- Reduced energy consumption of the PC dryer (3,000–3,500 kJ/kg H_2O) compared with conventional spray dryer (4,500–11,500 kJ/kg H_2O), fluidized bed dryer (4,500–9,000 kJ/kg H_2O), and flash dryers (4,500–9,000 kJ/kg H_2O) (Zbicinski 2002).
- High driving force of the process (about 700°C), which is limited only by the thermal sensitivity of the material to be dry (Zbicinski 2002).
- Decreased air pollutants emissions such as NO_x, CO, and soot of a PC burner compared with continuous combustion burner (Zbicinski, Benali, and Kudra 2002). Wang, Cui, and Sunada (2007) reported reduced concentrations of pollutants for Hypulcon pulse combustion spray dryer: NO concentration from ca. 20 to 40 ppm for heat input from ca. 12 to 19 kW (Wang, Cui, and Sunada 2007).

- Reduced drying time to below 1 s enables application of PC drying even for dewatering of heat-sensitive materials (Zbicinski 2002; "Ekonek" 2021). Application of PC drying has been reported for dewatering of the following agricultural products: apple pomace, whole eggs, fish meal as well as bio-materials: proteins, yeast, vitamins, antibiotics (Kudra 2008).
- Reduction of odor, volatile organic compound (VOC), and toxic contaminants emission (Meng, De Jong, and Kudra 2016).

One of the major drawbacks limiting wide application of PC technology in drying processes is high noise level, which in some valveless pulse combustors can reach 180 dB (A) (Zbicinski, Benali, and Kudra 2002). In the literature, the following methods of noise reduction in pulse combustors are reported:

- Reduction of operating pressure enables the decrease of sound pressure; however, lower pressure amplitudes also decrease heat and mass transfer rate, which increases the size of the dryer (Zbicinski, Benali, and Kudra 2002).
- The increase of the operation frequencies of pressure combustor allows for reduction of noise level. Researchers from Clean Energy Technology Inc. (Canada) reported reduction of noise level with simultaneous increase of operating frequencies from 60–70 cycles/min to 350–650 cycles/s applying new design of pulse combustor (Zhonghua and Mujumdar 2006).
- Application of two pulse combustors generating contrary sound waves (Zbicinski 2002).

Thus, the modern valved pulse combustors can operate at the noise level reduced to 75 dB (A) (Zbicinski, Benali, and Kudra 2002).

3.2.3 Application of Pulse Combustion Drying

Due to high flexibility, PCD may be applied for various types of products with different heat sensitivity and viscosity such as minerals and chemicals as well as food and agriculture products (Meng, De Jong, and Kudra 2016).

The examples of application of PCD with discussion of equipment, process parameters, and final product characteristics are provided in Table 3.2.

Wang, Cui, and Sunada (2007) applied commercially available PC spray dryer, Hypulcon (Pultech Corporation, Kobe, Japan) for drying of nitrendipine, a poorly water-soluble drug. As stated by the manufacturer, the Hypulcon drying system can achieve evaporation capacities from 0.5 to 6,500 kg/h on the H_2O basis with inlet temperatures up to 300°C and outlet temperatures about 60°C (Ohkawara Kakohki Co., Ltd. 2020). Nitrendipine has been dried applying PC and conventional spray dryer techniques for comparison. During PC drying, the following process parameters have been applied: drying chamber temperature of 60°C and feed rate of 8–13 mL/min, whereas during standard spray drying, inlet/outlet air temperature was 135°C and 75°C, feed rate was 10 mL/min, drying air flow rate was 0.6 mL/min, and atomizing air pressure was 50 kPa. The particles obtained by PCD were characterized by

TABLE 3.2

Application of Pulse Combustion Drying

Product	PC Dryer	Process Parameters	Product Characteristic and Observation	Reference
Nitrendipine with carriers (Aerosil, Tween80) Initial moisture content: 93%-98%	Hypulcon (Pultech Corporation, Kobe, Japan)	Drying chamber temperature: 60°C, Feed rate 8–13 mL/min	PSD after PCD: $D_{10} = 3.2$ µm, $D_{50} = 4.75$ µm, $D_{90} = 6.70$ µm, PSD after SD: $D_{10} = 3.2$ µm, $D_{50} = 6.2$ µm, $D_{90} = 11.6$ µm.	Wang, Cui, and Sunada (2007)
Ibuprofen with carriers (Kollidon 25, Kollidon 30, Kollidon VA64, Kollidon CL)	Hypulcon (Pultech Corporation, Kobe, Japan)	Drying temperature: 65°C, Feed rate: 8 mL/min Fuel flow rate: 30–35 L/h Combustion air feed rate: 900–1,000 L/h	PSD after PCD (Ibuprofen with Kollidon CL): $D_{10} = 16.1$ µm, $D_{50} = 47.4$ µm, $D_{90} = 119.2$ µm, PSD after SD: $D_{10} = 25.9$ µm, $D_{50} = 75.1$ µm, $D_{90} = 181.1$ µm. Dissolution rate constant after PCD: ca.0.02–0.08 min^{-1}, after SD: ca. 0.01–0.07 min^{-1}	Xu, Li, and Sunada (2007)
Egg white Initial moisture content: 86.96%	Pilot-plant PCD (Pulse Combustion System, USA) with Helmholtz-type PC	Inlet gas temperature: 326.6°C, Outlet gas temperature: 76°C Feed rate: 0.6 kg/min Fuel flow rate: 0.63 m³/min Heat release of the PC combustor: 24.32kW Energy consumption: 2,604 kJ/kg water evaporated Powder yield after PCD: 73.52%	PSD after PCD – $D_{50} = 20.15$ µm, Span – 2.71; PSD after SD – $D_{50} = 54.74$ µm, Span – 3.42; Morphology: after PCD – hollow structure of particles, after SD – dense solid structure; Foaming ability: after PCD – 26.3%, after SD – 38.0%, Foam stability: after PCD – 92.8 %, after SD – 96.5 %, Protein denaturation: 1.6% after PCD.	Wu et al. (2014); Rehkopf and Mirko (2017)

(Continued)

TABLE 3.2 (*Continued*)
Application of Pulse Combustion Drying

Product	PC Dryer	Process Parameters	Product Characteristic and Observation	Reference
Sewage sludge Initial moisture content of about 80% (wb) Sludge viscosity: 7,982x10³ mPa·S;	Laboratory pulse combustion dryer	PC frequency: 49 Hz, Heat load: 58 kW, LPG flow rate: 2 m³/h, Gas temperature in the tailpipe: 600°C–800°C, Pressure amplitudes: from −6 to 10 kPa, Sludge feed rate: 24 kg/h, Drying time: 0.5 s. Length of drying pipe: from 0.5 to 3 m.	PSD: 0.01 – 4 mm, Moisture content: 56%	Zhonghua et al. (2012)
Water, Maltose Viscosity: water – 0.001 Pa·S; Maltose –from 0.007 to 0.041 Pa·S	Laboratory pulse combustion dryer with Helmholtz-type PC	LPG fuel flow rate: 0.9 m³/h, PC frequency: 61–100 Hz, Heat load: 25 kW, Gas temperature in the tailpipe: 150°C, Pressure amplitudes: from 93.7 to 114.8 kPa, Liquid feed rate: from 35 to 62 L/h, Length of drying pipe: from 1 to 3 m.	Sauter mean diameter: Water – from 66 to 73 μm as feed rate changing from 35 to 62 L/h and from 99 to 66 μm as oscillating frequency changing from 61 to 100 Hz; Maltose – from 53 to 76 μm as viscosity changing from 0.007 to 0.041 Pa·S;	Xiao et al. (2008)
Refractory clay (SiO₂ 45%, Al₂O₃ 37% and others 18%) Initial moisture content: 0.32 kg/kg (dry basis)	Laboratory pulse combustion dryer with Helmholtz-type PC	LPG fuel flow rate: 0.8 m³/h, PC frequency: 39–75 Hz, Inlet gas temperature: 400°C, Exhaust gas temperature: 300°C Liquid feed rate: from 35 to 62 L/h, Length of drying pipe: from 1 to 4 m	Heat transfer coefficient for frequency 75 Hz – 287 W/m/K; The twofold increase of Nusselt number has been observed for an oscillating flow comparing to steady turbulent flow	Liu, Cao, and Lang (2001)

(Continued)

TABLE 3.2 (Continued)
Application of Pulse Combustion Drying

Product	PC Dryer	Process Parameters	Product Characteristic and Observation	Reference
Vegetable waste Initial moisture content: 66.06%	Pilot-plant PCD	Inlet temperature: 140°C	Moisture content: 25.20%, Protein content: PCD – 9.62%, oven drying 85°C – 8.87%, Ash content: PCD – 3.6%, oven drying 85°C – 2.84%,	San Martin, Ramos, and Zufia (2016)
ZnO	Pilot-plant installation for spray pyrolysis equipped with PC (Pultech, Kobe, Japan)	Fuel (propane) to oxidizer ratio: 1:26.6; Heat input: 1,000 kcal/h; Noise: 115 dB; Frequency: 1,000 Hz; Feed rates: 150, 250, or 500 mL/h; Atomizer: two-fluid nozzle or ultrasonic Furnace temperature: 300°C, 500°C, 800°C.	PSD: two-fluid nozzle $D_{50} = 56.8\,\mu m$, ultrasonic nozzle $D_{50} = 65.0\,\mu m$, for precursor feed rates: 150 mL/h – D_{50} = 102 μm, 250 mL/h – D_{50} =96.4 μm, 500 mL/h – D_{50} =72.3 μm.	Joni et al. (2009)
NaCl Initial moisture content: 90%–95%	Pilot-plant pulse combustion dryer with valved PC	Feed rate: 5 and 10 kg/h; Atomizing air feed rate: 7 kg/h;	Sauter mean diameter increases from ca. 60 to ca. 120 μm as distance from the atomizer increases from 10 to 60cm at the dryer axis. Axial velocity in the drying chamber is oscillating between ca. –0.2 and 0.5 m/s.	Strumillo et al. (1999)

higher solubility, narrow particle size distribution, smaller particles without agglom-erates compared with powders produced by classical spray drying.

Another drug poorly soluble in water, i.e., ibuprofen, has been dried by applying Hypulcon pulse combustion dryer at drying chamber temperature of 65°C, feed rate of 8 mL/min, propane feed rate of 30–35 L/h, and combustion air feed rate from 900 to 1,000 L/h. The authors compared the properties of ibuprofen particles obtained by PCD and standard spray drying. The particles produced by PCD had smaller diameters ($D_{50} = 47.4$ µm) and higher dissolution rate constant (ca. $0.02–0.08\,min^{-1}$) compared with standard spray drying ($D_{50} = 75.1$ µm and dissolution rate constant: ca. $0.01–0.07\,min^{-1}$).

Wu et al. (2014) reported successful PCD of egg white, the heat-sensitive prod-uct with increased protein content. The following PCD process parameters were applied: inlet/outlet air temperature of 326°C/76°C, feed rate of 0.6 kg/min. The authors showed that smaller particle size with narrow size distribution has been pro-duced after PCD ($D_{50} = 20.15$ µm, Span – 2.71) compared with standard spray drying ($D_{50} = 54.74$ µm, Span – 3.42). The protein denaturation in the egg white powder was only 1.6% after PCD.

PCD may be effectively applied for dewatering of high-viscosity materials such as sewage sludge with viscosity of $7,982 \times 10^3$ mPa·S (Zhonghua et al. 2012). Application of PC frequency of 49 Hz, pressure amplitudes in the range from −6 to 10 kPa as well as gas temperature in the tailpipe of 600°C–800°C resulted in the production of particles with the size varying from 0.01 to 4 mm and average drying time of 0.5 s.

Xiao et al. (2008) studied the effect of atomization parameters on the PDS of water and maltose droplets during PCD. Increase of feed rate from 35 to 62 L/h resulted in increase of water droplets, Sauter mean diameters (SMD) from 66 to 73 µm. The application of higher oscillating frequency (from 61 to 100 Hz) decreased the SMD from 99 to 66 µm, whereas increase of maltose solution viscosity from 0.007 to 0.041 Pa·S increased the SMD of maltose droplets from 53 to 76 µm.

PCD was proven to be suitable drying method for vegetable waste valorization and utilization as an animal feed (San Martin, Ramos, and Zufía 2016). Application of PCD at inlet temperature of 140°C allowed for decrease of vegetable waste mois-ture content from 66% to 25% and to produce final product with high protein content of 9.62%.

In the literature, the application of pulse combustors for spray drying of metal oxides SiO_2 and Al_2O_3 (Liu, Cao, and Lang 2001) as well as for spray pyrolysis of ZnO has also been reported (Joni et al. 2009).

3.3 FLAME SPRAY PYROLYSIS

3.3.1 History of Flame Spray Pyrolysis

Application of flame for the production of pigment particles from gases has been known for centuries, for example, manufacture of black ink for paintings in ancient China, India, Greece (Pratsinis 1998). Nowadays, flame technology is applied in the industrial scale for the production of carbon black widely applied as a reinforcing agent in automobile tires. Carbon black is produced by the combustion of fuel such as oil or gas at a high temperature in the range from 1,300°C to 1,500°C, the carbon black is collected as nanoparticles with sizes from 10 to 500 nm.

In 1984, Ulrih described the aerosol flame technology applied for the synthesis of SiO_2 nanoparticles from $SiCl_2$ vapor by oxidation reaction in the premixed flames (Ulrich 1984).

In the 1990s, aerosol flame technology became an attractive research field providing opportunity for the production of ceramic and metal oxide nanoparticles with desired properties for various applications. Kammler and Pratsinis (1999) scaled up the aerosol flame reactor applied for the production of SiO_2 and increased the production rate up to 130 g/h.

Later, Pratsinis and coworkers pioneered the development and investigation of flame spray pyrolysis (FSP) technique, where nanoparticles are synthesized from the precursor in the liquid phase, which gives advantage of application of wider spectrum of reactants compared with aerosol flame synthesis (Stark, Madler, and Pratsinis 2007). In 2002, Pratsinis research group from the Swiss Federal Institute of Technology (ETH Zurich) described controlled synthesis of SiO_2 (Mädler et al. 2002), Bi_2O_3 (Mädler and Pratsinis 2002), and CeO_2 (Mädler, Stark, and Pratsinis 2002) nanoparticles from the liquid precursors applying FSP.

3.3.2 Principle, Equipment, and Scale-Up of Flame Spray Pyrolysis

The FSP process covers combustion of the spray of metal-containing precursor solutions, which decomposes to yield single- and multicomponent nanoparticles (Hembram et al. 2013). FSP includes the following stages: precursor atomization into fine spray droplets, precursor evaporation into gas phase, precursor vapor oxidation in the flame, particle formation and subsequent particle agglomeration, and sintering (Teoh, Amal, and Mädler 2010). Figure 3.4 shows the combustion chamber for FSP consisting of an external-mixing pneumatic nozzle surrounded by pilot flames for ignition and stabilization of the flame (Buss, Noriler, and Fritsching 2020). Initially, the metal precursor is dissolved in the organic solvent such as ethanol, propanol, xylene, etc., and delivered to the two-fluid nozzle where oxygen is used as an atomizing medium. FSP allows for in situ synthesis of various types of nanoparticles with high specific surface area (SSA) controlled by the precursor feed rate and oxygen flow rate in the nozzle, precursor concentrations, flame height, and particle residence time (Kemmler et al. 2013).

For instance, formation of Zn_2SnO_4 nanoparticles by FSP from the zinc acetylacetonate ($C_{10}H_{14}O_4Zn$) and tin ethylhexanoate ($C_{16}H_{30}O_4Sn$) is described by the following steps:

- Evaporation and decomposition of zinc acetylacetonate to ZnO at 195°C:

$$C_{10}H_{14}O_4Zn + 12O_2 \rightarrow ZnO + 10CO_2 + 7H_2O \qquad (3.1)$$

- Evaporation and oxidation of tin ethylhexanoate to SnO_2 at 290°C:

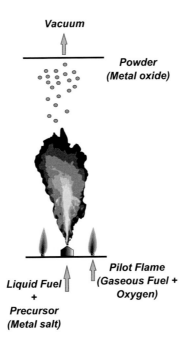

FIGURE 3.4 Combustion chamber for flame spray pyrolysis. (Based on Buss, Noriler, and Fritsching 2020.)

$$2C_{16}H_{30}O_4Sn + 45O_2 \rightarrow 2SnO_2 + 32CO_2 + 30H_2O \qquad (3.2)$$

- The maximum flame temperature applied in FSP process reaches 2,757°C, which exceeds the boiling points of ZnO (2,360°C) and SnO_2 (1,800°C–1,900°C); therefore, the following reaction of both species occurs in the gas phase:

$$2ZnO\,(g) + SnO_2\,(g) \rightarrow Zn_2SnO_4\,(g) \qquad (3.3)$$

- Zn_2SnO_4 nanoparticles are formed by nucleation and further grow by condensation and coalescence.
- Further Zn_2SnO_4 nanoparticle agglomeration (Kaewsiri et al. 2020).

The final nanoparticle quality parameters, i.e., SSA, might be controlled by the following FSP process parameters: type of precursor and solvent, concentration of precursor, feed rate of precursor and atomizing gas (oxygen or air) as well as parameters of supporting pilot flame: flow rate of fuel gas (commonly methane) and oxygen required for combustion as well as the flow rate of additional sheath O_2 stream (Mädler et al. 2002).

One of the first attempts to scale up the FSP process has been made by Mueller, Mädler, and Pratsinis (2003), who reported production of SiO_2 nanoparticles at the high production rate of 1.1 kg/h at the pilot plant FSP installation. Previously the nanoparticle production by FSP has been reported only at the production rate from ca. 9 (Mädler et al. 2002) to 400 g/h (Laine et al. 1999). To achieve high production rate, the authors applied two-fluid atomizing nozzle with external mixing of precursor/ethanol solution (from 5.55 to 33.3 mL/min) and atomizing O_2 stream (from 12.5 to 50 L/min). The nozzle was placed inside the two concentric stainless-steel tubes, which are used to supply the gaseous fuel, i.e., methane (2.0 L/min) and O_2 stream (4.5 L/min) needed to sustain the pilot flame (25 L/min). Additional sheath O_2 stream (25 L/min) has been supplied through third metal tube surrounding the burner as shown in Figure 3.5. To enhance the collection of produced nanoparticles, the four baghouse filters coated by polytetrafluoroethylene (PTFE) have been applied. The particle stream has been transferred from the FSP reactor to the baghouse filter by the suction fan (Mueller, Mädler, and Pratsinis 2003).

In 2013, Hembram et al. reported the production of ZnO nanorods by the FSP at the production rate of 3 kg/h. The authors applied microgear pump to supply 2–12 L/h of precursor solution to two-fluid nozzle, where it was atomized by 60–120 L/min

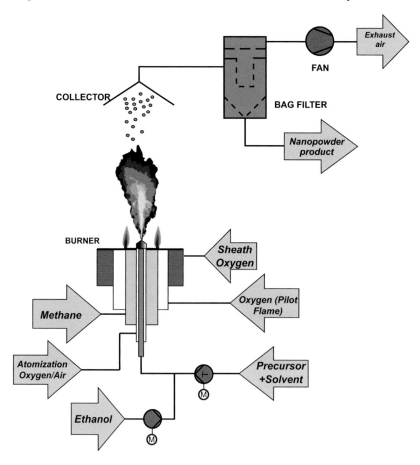

FIGURE 3.5   Schematic of FSP for SiO_2 synthesis at production rate of 1.1 kg/h. (Based on Mueller, Mädler, and Pratsinis 2003.)

of O_2. The combustion process was initiated by the pilot flame generated by 5 L/min of methane and 10 L/min of oxygen. To decrease the temperature of the stream, the additional co-flowing air has been supplied to the ring spacing between burner and reactor wall at the flow rate of 2,000–3,000 m³/h. The particles were collected by the baghouse filter, and additional HEPA filter had been installed on the exhaust air pipe to avoid nanoparticle release into the atmosphere (Hembram et al. 2013).

Scaling-up of FSP is difficult because high production rates could not be achieved by only increase of precursor flow rate since for higher precursor flow rates, the flame and particle residence time in the high temperature zone increase, which results in the undesirable increase of particle size. To avoid generation of large particles, the particle residence time in the high temperature zone may be decreased by the increase of atomization gas flow rate, which reduces the flame temperature in the reactor (Gröhn et al. 2014). Therefore, the constant ratio of atomization gas flow to precursor flow rate is recommended during the scale-up of FSP process (Mädler, Stark, and Pratsinis 2002; Gröhn et al. 2014).

3.3.3 ADVANTAGES AND DISADVANTAGES OF FSP

FSP advantages:

- easy to scale up, single-step process (Lee 2018),
- possibility to produce particles without impurities,
- easy collection of particles and control of particles quality parameters,
- high flexibility regarding the type of precursor and solvent,
- high efficiency of the process (Gonchikzhapov and Kasper 2020).

The following disadvantages of FSP might be pointed out:

- high investment costs (Gonchikzhapov and Kasper 2020);
- rigorous safety requirements caused by possible toxicity and harmful effect of nanoparticles, which involves application of personal protective equipment and advanced filter system (Weyell et al. 2020).

3.3.4 MECHANISM OF FLAME SPRAY PYROLYSIS

In the recent years, a large number of works on the mechanism of particles formation in FSP were published. Depending on the initial droplet size, precursor and solvent volatility, precursor decomposition temperature, and combustion enthalpy, the particle formation during FSP could be performed via two rotes: the gas-to-particle and the droplet-to-particle method (Liu et al. 2017).

The formation of particles via gas-to-particle route takes place if both solvent and precursor are highly volatile: after atomization, they both evaporate immediately and precursor oxidation reaction occurs in the gas phase (Gonchikzhapov and Kasper 2020). Gas-to-particle route results in homogeneous particles formation (Li et al. 2020), which is favorable for a variety of applications, for example, for the production of gas sensors.

Particles are produced by droplet-to-particle method if the solvent and precursor have low volatility, longer evaporation time, and particles are formed by droplets drying (Gonchikzhapov and Kasper 2020). The precursor decomposition occurs in the liquid phase within spray droplets and obtained nanoparticles grow by coagulation and sintering. The formation of particles via droplet-to-particles route results in different particles morphology: produced droplets have large particle size, dense, hollow, or porous structure (Li et al. 2020). For gas sensor application, high porosity of nanoparticles is unfavorable; however, it may be an advantage in catalysts production by FSP (Gonchikzhapov and Kasper 2020).

When solvent has higher volatility than precursor, the solvent may evaporate inside the droplet, whereas precursor decomposition at the droplet surface leads to a solid crust formation. Further droplet heating and pressure increase inside the crust may result in micro explosion of the droplet (Gonchikzhapov and Kasper 2020).

Li et al. (2020) analyzed single droplet combustion during SnO_2 production by FSP applying droplet generator and high-speed camera to record combustion process. The authors reported the large homogeneous SnO_2 particles formation via droplet micro explosion route. The small SnO_2 particles were obtained in the period between droplet inflation and micro explosion, where products of precursor decomposition are transferred to the flame from droplet surface. The increase of the precursor concentration in the initial solution reduces the time needed for droplet micro explosion, because of shorter time for solid crust formation. For example, the normalized time needed for micro explosion was ca. 0.48 $\mu s/\mu m^2$ for precursor concentration of 0.05 M (mol/L) and ca. 0.3 $\mu s/\mu m^2$ when precursor concentration has been increased to 1 M (mol/L) (Li et al. 2020).

Abram et al. (2019) investigated the production of Y_2O_3:Eu^{3+} nanoparticles by FSP at different flame temperatures. At low flame temperatures of 1,150 K, the FSP was carried out by the droplet-to-particle mechanism with formation of large, dense nanoparticles in the range from 10 to 100 nm. When flame temperature was increased up to 2,750 K, the particles with diameters from 4 to 5 nm were produced via gas-to-particle route.

3.3.5 APPLICATIONS OF FLAME SPRAY PYROLYSIS

In the recent years, the number of papers considering the utilization of FSP for the production of nanoparticles for different industrial applications has increased considerably. It's possible to apply a wide range of precursors (since almost every metal element can be utilized) as well as a variety of available solvents made FSP a versatile technique covering wide range of applications (Teoh, Amal, and Mädler 2010). In the next section, we focus on the relevant research on nanoparticles production by FSP for the following applications: catalysts, gas sensors, electrode materials for batteries, optical materials, and materials for medical applications.

3.3.5.1 Application of FSP for Catalysts Fabrication

Since in 1972, Fujishima and Honda had published a paper on photocatalytic decomposition of water into H_2 and O_2 on titanium dioxide electrodes (Fujishima and Honda

1972), the application of photocatalysis for degradation of environmental contamina-
tions acquired increasing attention of the scientific community (Guild et al. 2014).
Application of nanoparticles produced by FSP for photocatalysis includes CO_2 reduction (Xiong et al. 2017), reduction of NO_x as air pollutant (Giannouri et al. 2013),
removement of dyes from wastewater (Ajmal et al. 2014), and removement of VOCs
(Huang et al. 2016).

The photocatalytic activity of nanoparticles produced by FSP is determined by
two main particle characteristics: SSA and crystallinity, which could be controlled
by the following FSP process parameters: flame temperature, precursor concentration, and type of solvent applied (Guild et al. 2014).

TiO_2 catalyst combined with Cu^+ produced by FSP has been tested for photocatalytic reduction of CO_2 into methane (Xiong et al. 2020). The precursor, i.e., copper
(II) nitrate trihydrate and titanium butoxide, has been dissolved in ethanol with different molar concentration of Cu from 0.5 to 2 mol% and sprayed at a feed rate of
5 mL/min and atomizing O_2 flow rate of 5 L/min. Combination of Cu^+ with TiO_2
increased the SSA of nanoparticles from 47.24 to 105.62–93.58 m^2/g, which significantly increased the photocatalytic activity of Cu^+/TiO_2 catalyst compared with TiO_2
nanoparticles without additives (Xiong et al. 2020).

TiO_2 synthesized by FSP has been applied as a photocatalyst for degradation of
textile dye (Levafix Brilliant E-6BA) in the wastewater (Bahadori et al. 2020). The
authors applied titanium isopropoxide as a precursor and p-xylene and propionic acid
as a solvent at a flow rate of 2.5 mL/min and atomizing O_2 flow rate of 5 L/min. The
flame was sustained by pilot flame generated by methane combustion (CH_4 flow rate
of 0.5 L/min, O_2 flow rate of 1 L/min). Produced TiO_2 particles have been characterized by high SSA of 67.5 m^2/g and average particle size of 20 nm.

FSP titanium oxide catalyst showed lower photocatalytic activity for dye degradation compared with commercial TiO_2 (P25, Evonik) (Bahadori et al. 2020).

Jansanthea et al. (2019) applied FSP to obtain titanium oxide nanoparticles combined with gold for photocatalytic degradation of Rhodamine B dye. Two precursor
solutions – titanium isopropoxide and gold(III) chloride – have been dissolved in the
xylene solvent with Au concentrations of 0.50, 0.75, 1.00, and 3.00 mol%. The following process parameters were applied during FSP: precursor was sprayed at a flow rate
of 5 mL/min by atomizing O_2 at a flow rate of 5 L/min, the precursor combustion has
been supported by the pilot flame at CH_4 flow rate of 1.19 L/min and O_2 flow rate of
2.46 L/min (Chomkitichai et al. 2014). Combination of Au with TiO_2 nanoparticles
improved photocatalytic activity of catalyst, which increased with concentration of
Au in the precursor solution (Jansanthea et al. 2019).

Apart from synthesis of photocatalyst, FSP found an application for the production of various types of nanomaterial-based catalysts. For example, silicon
dioxide nanoparticles combined with palladium were obtained by FSP and tested
as a catalyst for removing the VOCs: toluene, benzene, and o-xylene mixture
(Guan et al. 2020). The authors applied precursor (palladium acetylacetonate,
hexamethyldisiloxane) and atomizing O_2 flow rate of 9.2 and 4.0 mL/min respectively and CH_4 (2.4 mL/min) and O_2 (4.0 mL/min) for pilot flame streams. The
synthesized Pd/TiO_2 nanoparticles had average particle size in the range from
3 to 10 nm and high SSA from 134 to 159 m^2/g, which resulted in high catalytic

activity for conversion of toluene even at low initial concentrations (110 ppm in air) (Guan et al. 2020).

Meng and Zhao (2020) used $CuO-TiO_2$ catalyst for low-temperature removal of toluene as a VOC air pollutant. The nanoparticles were produced by the FSP with tetrabutyl titanate and copper dinitrate as a precursor and ethanol as a solvent. The lower fraction of CuO in the $CuO-TiO_2$ resulted in higher SSA, i.e., for 2 wt.% CuO content in $CuO-TiO_2$ particles, the SSA was 98.96 m^2/g, and increase of CuO content to 50 wt.% resulted in lower SSA of 50.67 m^2/g. Samples with higher SSA showed enhanced catalyst activity (Meng and Zhao 2020).

FSP was applied for the synthesis of the nanomaterial-based catalyst for removal of 4-nitrophenol, which is widely utilized in the fabrication processes of anti-corrosion lubricants and pharmaceuticals (Psathas et al. 2020). Iron(III) acetylacetonate and bismuth(III) acetate precursors dissolved in the xylene and 2-ethylhexanoic acid have been decomposed in FSP reactor to yield $Bi_2Fe_4O_9$ and $BiFeO_3$ catalyst nanoparticles. The produced catalyst was characterized by particle size from 60 to 230 nm and SSA from 2.8 to 8.9 m^2/g (Psathas et al. 2020).

Nanoparticles produced by FSP may be also applied as a catalyst for conversion of biomass particularly of lignocellulosic nature into valuable platform chemicals. For example, Beh et al. (2020) synthesized $SiO_2-Al_2O_3$ and $SiO_2-Al_2O_3$ phosphate nanoparticle catalyst for transformation of glucose coming from biomass conversion into levulinic acid, i.e., valuable platform chemical. The authors reported the decrease of SSA from ca. 400 to 200 m^2/g with increase of Al fraction in the $SiO_2-Al_2O_3$ particles (Beh et al. 2020). Silica alumina and silica alumina phosphate nanoparticles produced by FSP showed high catalytic performance in conversion of glucose to levulinic acid with yield 40%, whereas application of conventional catalysts resulted in lower yield of levulinic acid, i.e., 17% and 21% (Beh et al. 2020).

ZnO nanoparticles obtained by FSP have been tested as a catalyst for conversion of CO_2 into syngas (mixture of H_2 and CO) by electrochemical reduction reaction (Daiyan et al. 2020). The SSA of nanoparticles was obtained by applying different precursor (zinc 2-ethylhexanoate dissolved in xylene) flow rate: for 5 mL/min – 65 m^2/g, for 7 mL/min – 55 m^2/g, and for 9 mL/min – 69.3 m^2/g. Particle size increased with increase of precursor flow rate: for 5 mL/min – 16.8 nm, for 7 mL/min – 18.3 nm, and for 9 mL/min – 21.6 nm (Daiyan et al. 2020).

Table 3.3 summarizes the discussed papers on FSP synthesis of the catalyst for environmentally friendly applications. The table contains information on the type of nanoparticles obtained by FSP, type of precursors and solvents applied, the target application of nanoparticles, the FSP process conditions, and data on the key nanoparticles characteristics: particle size and SSA.

3.3.5.2 Application of FSP for Sensors Fabrication

Due to numerous advantages, FSP process has found applications for manufacture of nanoscale materials, i.e., semiconducting metal oxides for gas sensors. The major problem of application of metal oxides in the gas sensors is poor selectivity and high susceptibility to ambient humidity (Kemmler et al. 2013). The flexibility in selection of precursors, additives, and solvents provided by FSP enables to enhance the sensing properties toward the specific gas. High crystallinity, external surface area,

TABLE 3.3

Application of FSP for Catalysts Fabrication

Product	Precursor	Application	Process Parameters	Product Characteristic	Reference
Au/TiO_2	Titanium isopropoxide, gold(III) chloride, Solvent: xylene, Au concentration: 0.50, 0.75, 1.00, and 3.00 mol%	Photocatalytic degradation of Rhodamine B dye	Precursor flow rate: 5 mL/min, Atomizing O_2 flow rate: 5 L/min Pilot flame: CH_4 (1.19 L/min), O_2 (2.46 L/min).	Particle size: <100 nm, particle size increased with Au concentration in the precursor solution, the photocatalytic activity increased with increase in Au concentration (up to 1.00 mol%).	Chomkitichai et al. (2014); Jansanthea et al. (2019)
$Bi_2Fe_4O_9$, $BiFeO_3$	Iron(III) acetylacetonate, Bismuth(III) acetate, Solvents: xylene, 2-Ethylhexanoic acid	Catalyst for removal of 4-nitrophenol	Precursor flow rate: 3 mL/ min, Atomizing O_2 flow rate: 7 L/min Pilot flame: CH_4 (2.5 L/min), O_2 (5 L/min)	SSA: 2.8–8.9 m^2/g, Particle size: from 60 to 230 nm	Psathas et al. (2020)
Cu^+/TiO_2	$Cu(NO_3)_2 \cdot 3H_2O$, titanium butoxide. Solvent: ethanol Molar concentration of Cu/ (Cu + Ti): 0.5, 1, and 2 mol%	Photocatalyst for CO_2 reduction by water	Precursor flow rate: 5 mL/min, Atomizing O_2 flow rate: 5 L/min	Combination of Cu with TiO_2 nanoparticles increased SSA from 47.24 to 105.62–93.58 m^2/g	Xiong et al. (2020)
CuO/ZrO_2	copper acetate monohydrate, zirconyl 2-ethylhexanoate, Solvent: 2-ethylhexanoic acid, methanol. Metal concentration: 0.2 M	Catalyst for hydrogenation of CO_2 to methanol	Precursor flow rate: 1–10 mL/min, Atomizing O_2 flow rate: 8 L/min Pilot flame: CH_4 (1.5 L/min), O_2 (3.2 L/min). Additional O_2 flow rate: 5 L/min	The crystallite size of ZrO_2 reduced from 6.9 to 2.6 nm as feed rate decreased from 10 to 1 mL/min.	Fujiwara et al. (2019)
$CuO-TiO_2$	Tetrabutyl titanate, copper dinitrate. Solvent: Ethanol Cu/Ti mass ratio: from 2 to 50 wt.% (content of CuO in $CuO-TiO_2$ particles).	Catalyst for removing of toluene	Precursor flow rate: 5 mL/ min, Precursor atomization pressure: 1.5 bar, Atomizing O_2 flow rate: 5 L/min Pilot flame: CH_4 (0.75 L/min), O_2 (1.5 mL/min)	BET diameter: 15.1–22.3 nm SSA: 50.67–98.96 m^2/g	Meng and Zhao (2020)

(Continued)

TABLE 3.3 (*Continued*)
Application of FSP for Catalysts Fabrication

Product	Precursor	Application	Process Parameters	Product Characteristic	Reference
Pd-SiO$_2$	Palladium acetylacetonate, hexamethyldisiloxane (mass ratio of Pd/Si was 0.5%)	Catalyst for removing the volatile organic compounds (VOC): toluene, benzene and o-xylene mixture	Precursor flow rate: 9.2 mL/min, Atomizing O$_2$ flow rate: 4.0 mL/min Pilot flame: CH$_4$ (2.4 mL/min), O$_2$ (4.0 mL/min)	Average particle size: ca. 3–10 nm SSA: ca. 134–159 m^2/g	Guan et al. (2020)
SiO$_2$-Al$_2$O$_3$, SiO$_2$-Al$_2$O$_3$ phosphate	Tetraethyl orthosilicate, aluminum tri-*sec*-butoxide, tributyl phosphate. Solvent: xylene	Catalyst for conversion of glucose to levulinic acid, one of the valuable platform chemicals	Precursor flow rate: 1–5 mL/min, Atomizing O$_2$ flow rate: 5 L/min Atomizing pressure: 1.5 bar Pilot flame: CH$_4$ (1.5 L/min), O$_2$ (3.2 L/min). Additional O$_2$ flow rate: 5 L/min	SSA decreased from ca. 400 to ca. 200 m^2/g with increase of Al fraction in the Al+Si particles.	Beh et al. (2020)
TiO$_2$	Titanium Isopropoxide, Solvent: p-xylene, propionic acid Precursor concentration 0.4 M	Photocatalyst for degradation of textile dye in the wastewater	Precursor flow rate: 2.5 mL/min, Atomizing O$_2$ flow rate: 5 L/min Atomizing pressure: 1.5 bar Pilot flame: CH$_4$ (0.5 L/min), O$_2$ (1 L/min).	SSA: 67.5 m^2/g Particle size: 20 nm	Bahadori et al. (2020)
ZnO	Zinc 2-ethylhexanoate, Solvent: xylene Metal concentration: 0.5 M	Catalyst in conversion of CO$_2$ into syngas by electrochemical reduction reaction.	Precursor flow rate: 5, 7 and 9 mL/min, Atomizing O$_2$ flow rate: 5 L/min Atomizing pressure: 1.5 bar Pilot flame: CH$_4$ (1.5 L/min), O$_2$ (3.2 L/min). Additional O$_2$ flow rate: 5 L/min	SSA for different precursor feed rate: 5 mL/min – 65 m^2/g, 7 mL/min – 55 m^2/g and 9 mL/min – 69.3 m^2/g, Particle size: 5 mL/min – 16.8 nm, 7 mL/min – 18.3 nm and 9 mL/min – 21.6 nm	Daiyan et al. (2020)

and thermal stability are the key requirements that should be met by the metal oxide nanoparticles to be applied as a gas sensor. In terms of gas sensing applications, the precursor composition and type of solvent applied are the main process parameters affecting the quality of produced nanoparticles (Kemmler et al. 2013).

FSP was applied to produce $LaFeO_3$ as well as Ag-loaded $LaFeO_3$ nanoparticles utilized as a material for gas sensor dedicated to acetylene detection (Sukee et al. 2020). The following precursors were applied: $La(NO_3)_3 \cdot H_2O$, $Fe(NO_3)_3 \cdot 9H_2O$, $Ag(NO_3)$, and solvents: acetonitrile and methanol under the following process conditions: precursor flow rate: 5 mL/min, atomizing O_2 flow rate: 5 L/min, whereas pilot flame was sustained by CH_4 (1.19 L/min) burned in O_2 stream (2.46 L/min). The diameters of produced nanoparticles were in range from 50 to 500 nm. $LaFeO_3$ sensor for acetylene detection produced by FSP has improved detection properties compared with similar sensor obtained by conventional sol–gel method: higher selectivity at lower acetylene concentrations (Sukee et al. 2020).

The mechanism of nanoparticles formation of tin oxide (SnO_2) for gas sensing application has been evaluated experimentally by single droplet combustion test and by FSP. The effect of Sn concentration in the precursor solution (tin(II) 2-ethylhexanoate, xylene) on particle diameters and SSA has been analyzed. Particle diameters increased from 5.2 to 11.1 nm, whereas SSA decreased from ca. 160 to 80 m^2/g when the concentration of the Sn in the precursor solution has been increased from 0.05 to 1 mol/L.

To enhance the performance of gas sensor, the SnO_2 nanoparticles were combined with Pd and have been tested for the detection of acetone, ethanol, and CO. The authors applied tin(II)-ethylhexanoate and palladium(II)-acetylacetonate precursor dissolved in xylene at the standard process parameters: precursor flow rate 5 mL/min and atomizing O_2 flow rate 5 L/min. The produced Pd/SnO_2 nanoparticles were able to detect low concentrations of acetone of ca. 5 ppb at high relative humidity of 50% (Pineau et al. 2020).

Kaewsiri et al. (2020) combined two metal oxides SnO_2 and ZnO_2 with high gas-sensing activity and produced Zn_2SnO_4 by FSP technique. Obtained Zn_2SnO_4 nanoparticles due to high SSA (BET): 78.8 m^2/g and low particles diameter (BET): 11.8 nm showed high selectivity in formic acid detection within the concentrations in range from 16.5 to 1,000 ppm.

To improve sensing properties, SnO_2 nanoparticles were combined with silver oxide via FSP of mixture of tin(II) 2-ethylhexanoate and silver nitrate precursors dissolved in xylene and acetonitrile. The authors evaluated the effect of silver concentration in the produced nanoparticles on SSA of powders and particle size. SSA (BET) increases from 52.4 to 77.3 m^2/g, and particle diameters (BET) decrease from 11.5 to 7.6 nm as the content of Ag increases from 0 to 0.2 wt.% (Khamfoo et al. 2020). The SnO_2 nanoparticles with addition of Ag at optimal concentration (0.2 wt.%) showed high selectivity in formaldehyde detection at concentrations in the range from 495 to 2,000 ppm.

In addition to SnO_2, another metal oxide may be applied as a material for gas sensors. For example, Fe_3O_4 obtained via FSP has been tested as a material for optical gas sensor for O_2 detection. Fe_3O_4 nanoparticles with average particle size of

60 nm have been produced by applying iron(III) 2,4-pentanedionate as a precursor dissolved in ethanol and acetic acid (Oguzlar 2020).

Tungsten(VI) ethoxide, bismuth(III) nitrate pentahydrate has been applied to obtain Bi_2WO_6 nanoparticles with high sensing properties for hydrogen sulfide H_2S detection (Bunpang et al. 2019). The authors used standard FSP process parameters: precursor flow rate of 5 mL/min and atomizing O_2 flow rate of 5 L/m to produce nanoparticles with SSA of about $20 m^2/g$ and particle size in the range from 5 to 15 nm. Combination of Bi_2WO_6 nanoparticles with reduced graphene oxide resulted in high sensor response and selectivity. The sensor was able to detect low concentration of H_2S (10 ppm) selectively in the presence of other gases such as H_2, CH_4, NO, NO_2, C_7H_8, CH_2O, C_8H_{10}, C_6H_6, C_3H_6O, CH_3OH, C_2H_5OH, $C_3H_6O_2$, $C_3H_6O_3$, $C_4H_8O_2$, CH_3COOH, C_4H_9COOH, and HCOOH.

Table 3.4 includes the data on the type of nanoparticles obtained by FSP for gas sensor applications, type of precursors and solvents applied, the target substance, which could be detected by sensor, the FSP process parameters, and data on the key nanoparticles characteristics: particle size and SSA.

3.3.5.3   Application of FSP for Electrode Material Fabrication

Metal oxide nanoparticles fabricated by FSP have found wide application as electrode materials used in fuel cells and electric vehicles and as an element in different types of batteries: lithium-ion, sodium-ion, and magnesium ion batteries. Nanoparticles for electrode application should fulfill the following requirements: high surface area and conductivity, well-defined and stable pore structure, and high corrosion resistance (Dahl et al. 2015). The most important FSP process parameters, which affect the electrochemical performance of nanoparticles for electrode materials, are the type and concentration of precursor and solvent, precursor, and atomization air flow rate.

The most recent papers describing applications of FSP for production of electrode materials are discussed below.

To obtain Ni-based $LiNi_{0.815}Co_{0.15}Al_{0.035}O_2$ cathode material for application in electric vehicles, Yudha et al. (2020) used urea as a fuel additive in FSP. Three different precursors $Ni(NO_3)_2 \cdot 6H_2O$, $Co(NO_3)_2 \cdot 6H_2O$, $Al(NO_3)_3 \cdot 9H_2O$ have been mixed and atomized at the precursor and atomizing air flow rate of 0.5 mL/min and 4 L/h respectively, the combustion process has been sustained by LPG fuel supplied at flow rate of 1 L/h. The applied process parameters allowed for production of nanoparticles within narrow particle size distribution in the range from 1 to 25 μm. The electrochemical characteristics of the produced cathode material have been affected by the precursor solution concentration; for instance, the cathode produced from the 1 M precursor solution concentration had the highest initial specific discharge capacity of 155 mAh/g and highest capacity retention of 92%. The $LiNi_{0.815}Co_{0.15}Al_{0.035}O_2$ cathode obtained by FSP has enhanced physicochemical and electrochemical properties compared with commercial Ni-cathode available on the market (Yudha et al. 2020).

The application of FSP-based nanoparticles as a material for sodium-ion batteries is currently an attractive research field. The sodium-ion batteries show similar energy-storage capacity as lithium-ion batteries; however, compared with Li-ion system, Na-ion batteries have larger ionic radius and higher reduction potential, which

TABLE 3.4

Application of FSP for Sensors Fabrication

Product	Precursor	Application	Process Parameters	Product Characteristic	Reference
$LaFeO_3$, Ag-loaded $LaFeO_3$ (0.1 and 1 wt.%)	$La(NO_3)_3 \cdot H_2O$, $Fe(NO_3)_3 \cdot 9H_2O$, $Ag(NO_3)$ solved in acetonitrile, methanol solvent	Sensor for acetylene detection	Precursor flow rate: 5 mL/min, Atomizing O_2 flow rate: 5 L/min Pilot flame: CH_4 (1.19 L/min), O_2 (2.46 L/min) Additional O_2 flow rate: 3.92 L/min	Particles size: 50–500 nm Sensing for ethylene and acetylene: 5–50 ppm in dry and humid air at temperature from 150°C to 300°C	Sukee et al. (2020)
SnO_2	Tin(II) 2-ethylhexanoate, Solvent: xylene Sn concentration in precursor solution: 0.05–1 mol/L	Application for gas sensors	Precursor flow rate: 5 mL/min, Atomizing O_2 flow rate: 5 L/min Pilot flame: CH_4 (1.5 L/min), O_2 (3.2 L/min)	Particles diameters increased from 5.2 to 11.1 nm when concentration of the Sn in the precursor solution has been increased from 0.05 to 1 mol/L. SSA decreased from ca. 160 to 80 m²/g with increase of Sn concentration in the precursor solution from 0.05 to 1 mol/L.	Li et al. (2020)
Pd-containing SnO_2	tin(II)-ethylhexanoate, palladium(II)-acetylacetonate Solvent: xylene Pd content: 0–3 mol%	Application for gas sensors for detection of acetone, ethanol, CO.	Precursor flow rate: 5 mL/min, Atomizing O_2 flow rate: 5 L/min Pilot flame: CH_4 (1.25 L/min), O_2 (3.25 L/min) Additional O_2 flow rate: 5 L/min	Particle size: 15 – 21 nm	Pineau et al. (2020)
Zn_2SnO_4	Zinc(II) acetylacetonate ($C_{10}H_{14}O_4Zn$), Tin(II) 2-ethylhexanoate ($C_{16}H_{30}O_4Sn$), Solvents: 30 vol% methanol, 70 vol% xylene	Gas sensor for formic acid	Precursor flow rate: 5 mL/ min, Atomizing O_2 flow rate: 5 L/min	SSA (BET): 78.8 m²/g, Particles diameter (BET): 11.8 nm.	Kaewsiri et al. (2020)

(Continued)

TABLE 3.4 (*Continued*)
Application of FSP for Sensors Fabrication

Product	Precursor	Application	Process Parameters	Product Characteristic	Reference
AgO_x-doped SnO_2 nanoparticles	Tin(II)2-ethylhexanoate $[CH_3(CH_2)_3CH(C_2H_5)CO_2]_2Sn$, silver nitrate $AgNO_3$, Solvents: xylene $(C_6H_4(CH_3)_2$, acetonitrile CH_3CN Metal concentration: 0.5 M	Gas sensor for formaldehyde (HCHO)	Precursor flow rate: 5 mL/min, Atomizing O_2 flow rate: 5 L/min Pilot flame: CH_4 (1.19 L/min), O_2 (2.45 L/min), Additional O_2 flow rate: 3.95 L/min	SSA (BET) increases from 52.4 to 77.3 m^2/g as the content of Ag increases from 0 to 0.2 wt.%. Particles diameters (BET) decreases from 11.5 to 7.6 nm as the Ag content rises from 0 to 0.2 wt.%.	Khamfoo et al. (2020)
Fe_3O_4	Iron(III) 2,4-pentanedionate, Solvent: ethanol, acetic acid	Optical gas sensor for O_2	FSP equipment: Np10 (Tethis, Milan, Italy)	Average particles size: 60 nm.	Oguzlar (2020)
Bi_2WO_6	Tungsten(VI) ethoxide, bismuth (III) nitrate pentahydrate, Solvent: acetic acid, ethanol	Gas sensor for hydrogen sulfide (H_2S)	Precursor flow rate: 5 mL/min, Atomizing O_2 flow rate: 5 L/min.	SSA: ca. 20 m^2/g, Particles size: 5–15 nm	Bunpang et al. (2019)

improve the electrochemical properties of Na-ion batteries (Kim, Kim, and Kang 2020).

Magnesium-ion batteries as an alternative for conventional Li-ion batteries possess a range of advantages: the raw materials for Mg-ion batteries are safe and inexpensive, whereas their specific and volumetric capacity is high. Thus, $Mg_{0.5(1+x)}Fe_xZr_{2-x}(PO_4)_3$ nanoparticles produced by FSP for application in Mg-ion batteries were characterized by high SSA of $23\,m^2/g$, small average particle size of $80\,nm$, and improved ionic conductivity (Liu et al. 2020).

Zhang et al. (2020) produced Ni-based cathode material for Li-ion batteries, i.e., $LiNi_{0.8}Co_{0.15}Al_{0.05}O_2$ particles, applying FSP and conventional spray drying process. The particles from FSP had slightly higher SSA ($3.5\,m^2/g$), then after classical spray drying ($2.5\,m^2/g$), and lower particle size: from 10 to $20\,\mu m$ after FSP and from 20 to $40\,\mu m$ after conventional spray drying. Higher SSA of particles obtained by FSP resulted in improved electrochemical properties: particles after FSP have an initial discharge capacity of $200.2\,mAh/g$ at the rate of $0.1\,C$, whereas the sample produced by standard spray drying has an initial discharge capacity of $185.1\,mAh/g$ at the same rate.

Table 3.5 shows the examples of nanoparticles produced by FSP for application as an electrode material.

3.3.5.4 Application of FSP for Optical Materials Fabrication

Flexibility of FSP provided by possible application of wide range of inexpensive precursors opens a new perspective for production of nanoparticles for optical materials. Depending on the type of optical material, the requirements to nanoparticles produced by FSP might differ; however, the most common requirements are narrow particle size distribution, non-hollow and homogeneous particle structure.

Yttrium aluminum oxide nanoparticles are commonly applied composite for optical materials due to high optical clarity and mechanical strength (Wei et al. 2019). There are a few papers describing the application of FSP for production of yttrium aluminum oxide nanocomposites.

$Y_3Al_5O_{12}$ nanoparticles for application in laser technology have been produced by FSP applying low-cost raw materials, i.e., metal nitrate precursors ($Y(NO_3)_3 \cdot 6H_2O$, $Al(NO_3)_3 \cdot 9H_2O$) and different types of solvents: anhydrous ethanol, anhydrous butanol, and mixture of 2-ethylhexanoic acid and ethanol (Wei et al. 2019). The researchers found that the type of solvent applied for FSP affected the mechanism of nanoparticles formation. Application of ethylhexanoic acid as a solvent involved gas-to-particles route and resulted in nanoparticle size ca. $10\,nm$, whereas application of butanol and ethanol as a solvent promotes droplet-to-particle route of nanoparticles formation. Such effect of ethylhexanoic acid application has been explained by the synthesis of intermediate product 2-ethylhexanoates with low boiling point, which resulted in fast evaporation of droplets and further gas-to-particle conversion route. This phenomenon resulted in the formation of homogeneous particles with narrow particle size distribution, which was not observed when ethanol or butanol has been applied as a solvent.

To improve the optical properties, yttrium aluminum oxide nanoparticles might be combined with other metals such as neodymium (Nd) or europium (Eu).

TABLE 3.5

Application of FSP for Electrode Material Fabrication

Product	Precursor	Application	Process parameters	Product characteristic	Reference
$LiNi_{0.815}Co_{0.15}Al_{0.035}O_2$	$Ni(NO_3)_2 \cdot 6H_2O$, $Co(NO_3)_2 \cdot 6H_2O$, $Al(NO_3)_3 \cdot 9H_2O$, urea $CO(NH_2)_2$ applied as a fuel additive (60 mg/mL), Precursor solution concentration from 0.5 to 2.0 M	Cathode for application in electric vehicles	Precursor flow rate: 0.5 mL/min, Atomizing air flow rate: 4 L/h Fuel: LPG (1 L/h)	Particle size: from ca. 1 to 25 μm	Yudha et al. (2020)
MoO_3, MoS_2, $MoS_2@NC$	Ammonium molybdate tetrahydrate and water as a solvent	Anode material for sodium-ion batteries	Precursor flow rate: 1.5 mL/min, O_2 flow rate: 10 L/min Pilot flame: C_3H_8 (5 L/min), O_2 (40 mL/min) Production rate: 2.7 g/h Ultrasonic nebulizer frequency 1.7 MHz	SSA (BET): 6.7 and 10.7 m²/g.	Kim, Kim, and Kang (2020)
$Mg_{0.5(1+x)}Fe_xZr_{2-x}(PO_4)_3$ $x = 0, 0.1, 0.2$	$Mg(CH_3CH_2COO)_2$, $Zr[(CH_3)_2CHCO_2]_2(OH)_2$, $Fe(O_2CCH_2CH_3)_3$ Solvent: ethanol	Electrode material for magnesium-ion batteries	nd	SSA: 23 m²/g, Average particle size: 80 nm	Liu et al. (2020)
$LiNi_{0.8}Co_{0.15}Al_{0.05}O_2$	$LiCH_3COO \cdot 2H_2O$, $Ni(CH_3COO)_2 \cdot 4H_2O$, $(CH_3COO)_2Co \cdot 4H_2O$ and $Al (NO_3)_3 \cdot 9H_2O$ Solvent: glycerol	Cathode material for Li-ion batteries	Powders were produced by flame assisted spray drying and conventional spray drying (SP), Temperature in the drying zone: 270°C	SSA: FSP – 3.5 m²/g, SD – 2.5 m²/g. Particle size: FSP – from 10 to 20 μm, SD – from 20 to 40 μm.	Zhang et al. (2020)

For example, Liu et al. (2020) applied FSP to produce $Y_4Al_2O_9$ and $Y_{4-x}Eu_xAl_2O_9$ ($x = 0.05-1.0$) nanoparticles designated for application as a photoluminescence emitting elements. To produce the nanoparticles, the mixture of the following precursors: $Y(NO_3)_3 \cdot 6H_2O$, $(Y(OC_4H_9)_3$, $Al[OCH(CH_3)C_2H_5]_3$, $EuC_{24}H_{45}O_6$, and solvents: tetrahydrofuran, 2-ethylhexanoic acid was used. The authors compared the SSA and particle diameters for both yttrium aluminum oxide and yttrium aluminum/europium oxide nanoparticles. Addition of Eu resulted in the increase of SSA and decrease of particle size: for $Y_4Al_2O_9$ nanoparticles, SSA was 84.4 m²/g and particle diameter (BET) was 16.1 nm, for $Y_{4-x}Eu_xAl_2O_9$ ($x = 0.05-1.0$), SSA was is in the range from 96.7 to 117.8 m²/g and particle diameters from 10.7 to 13.3 nm (Liu et al. 2020). For $Y_{4-x}Eu_xAl_2O_9$ nanoparticles, the increase of Eu concentration from $x = 0.05$ to $x = 0.5$ resulted in higher photoluminescence intensity, further increase in Eu concentration to from $x = 0.7$ to $x = 1.0$ resulted in reduction of photoluminescence intensity.

To obtain nanoparticles, which could be applied as a material for high-power laser technology, Sakar et al. (2020) combined yttrium aluminum oxide with Nd in the single-step FSP using metal nitrates ($Y(NO_3)_3 \cdot 6H_2O$, $Al(NO_3)$ $9H_2O$, $Nd(NO_3)_3 \cdot 6H_2O$) as a precursor and ethanol as a solvent. The produced particles with chemical formula $Nd_xY_{3-x}Al_5O_{12}$ ($x = 0, 1, 3, 5$) had particles size in the range from 200 to 1,000 nm. Addition of Nd to yttrium aluminum oxide improved the luminescence properties of nanoparticles; however, application of Nd concentration above 1% results in lower luminescence intensity (Sakar et al. 2020).

The examples of application of FSP for production of optical materials are summarized in Table 3.6.

3.3.5.5 Application of FSP for Medical Materials Fabrication

One of the important fields of applications of FSP-based nanoparticles is the production of nanocarriers for drug delivery. There are several substances that could be applied as a medical treatment in certain diseases; however, their application may be limited due to problems during delivery of such substances to the human body. As an example, biological drugs such as proteins, antibodies, peptides, which are difficult to deliver via oral route or due to their susceptibility to enzymatic degradation in tissues. To enhance application of such drugs, the development of drug delivery system based on the nanocarriers is needed. Nanocarriers for drug delivery systems should possess the following properties: biocompatibility, low toxicity, and high SSA to ensure high drug content loading. Thus, calcium phosphate nanoparticles produced by FSP might be applied as nanocarriers for delivery of biological drugs, i.e., bovine serum albumin and bradykinin (Tsikourkitoudi et al. 2020). For example, CaP nanoparticles were produced at precursor flow rate: 8 and 3 mL/min and atomizing O_2 flow rate: 3 and 8 L/min. The produced nanoparticles were characterized by high SSA 73 and 246 m²/g and small particle diameters (BET): 26 and 8 nm (Tsikourkitoudi et al. 2020).

Ataol et al. (2015) proposed to apply calcium phosphate nanoparticles obtained by FSP as a bone substitute and bone filling cement. To be applied as a biomaterial, calcium phosphate nanoparticles should be biodegradable, biocompatible, and have large surface-to-volume ratio to provide area for crystal grow and further biomineralization. Previous studies showed that calcium phosphate in amorphous

TABLE 3.6

Application of FSP for Optical Materials Fabrication

Product	Precursor	Application	Process Parameters	Product Characteristic	Reference
$Nd_xY_{3-x}Al_5O_{12}$	$Y(NO_3)_3 \cdot 6H_2O$, $Al(NO_3)_3 \cdot 9H_2O$, $Nd(NO_3)_3 \cdot H_2O$, Solvent: Ethanol	Crystals for high-power laser technology and solid-state laser devises	Precursor flow rate: 5 mL/min, Pilot flame: CH_4 (1.5 L/min), O_2 (3 L/min) Additional O_2 flow rate: 3.92 L/min	Particle size: 200–1,100 nm	Sakar et al. (2020)
$Y_4Al_2O_9$, $Y_{4-x}Eu_xAl_2O_9$ ($x = 0.05$–1.0)	$Y(NO_3)_3 \cdot 6H_2O$, $(Y(OC_4H_9)_3$, $Al[OCH(CH_3)C_2H_5]_3$, $EuC_{24}H_{45}O_6$ Solvents: Tetrahydrofuran, 2-ethylhexanoic acid Metal concentration: 0.1 or 0.5 M	Photoluminescence emitting source	Precursor flow rate: 3 and 7 mL/min, Atomizing O_2 flow rate: from 3 to 7 L/min Pilot flame: CH_4 (1.5 L/min), O_2 (3.2 L/min)	SSA: $Y_4Al_2O_9$ – 84.4 m²/g, $Y_{4-x}Eu_xAl_2O_9$ ($x = 0.05$–1.0) is in the range from 96.7 to 117.8 m²/g, Particles diameter (BET): $Y_4Al_2O_9$ – 16.1 nm, $Y_{4-x}Eu_xAl_2O_9$ ($x = 0.05$–1.0) is in the range from 10.7 to 13.3 nm.	Liu et al. (2020)
$Y_3Al_5O_{12}$	Yttrium nitrate hexahydrate $(Y(NO_3)_3 \cdot 6H_2O)$, aluminum nitrate nonahydrate $(Al(NO_3)_3 \cdot 9H_2O)$, Solvents: Anhydrous ethanol, Anhydrous butanol, Mixture of 2-ethylhexanoic acid and ethanol. The Y/Al ratio – 3:5; concentration of Y^{3+} is 0.15 mol/L	Laser technology	Atomizing N_2 flow rate: 0, 0.1, 0.25 L/min Pilot flame: CH_4 (0.25, 0.15, 0 L/min), O_2 (0.55 L/min)	Application of ethylhexanoic acid as a solvent involves gas-to-particles route for nanoparticles formation with size ca. 10 nm, whereas application of butanol and ethanol as a solvent promote droplet-to-particle route of nanoparticles formation.	Wei et al. (2019)

form shows enhanced bioactive properties (Sarkar et al. 2001). Amorphous calcium phosphate nanoparticles have been produced by FSP using calcium acetate hydrate and tributyl phosphate as a precursors and propionic acid as a solvent (Ataol et al. 2015). The increase in Ca/P ratio from 1.27 to 2.29 resulted in the increase of SSA of nanoparticles from ca. 40 to ca. $50\,m^2/g$, however, with no effect on the average particles diameter, which was about $23\,nm$ for all samples. The FSP-based nanoparticles showed high biocompatibility during in vitro studies.

Another possible application of FSP is production of nanoparticles for gas sensors, which could be applied in medicine. For example, FSP-based tungsten oxide (WO_3) nanoparticles combined with silica (Si) have been applied as a material for breath acetone sensor for regulation of ketosis during ketogenic diet (Güntner et al. 2018). The obtained Si/WO_3 particles had average particle size (BET) about $12\,nm$ and showed high sensing response and selectivity in detection of acetone at low concentrations up to ca. $66\,ppm$.

Nanoparticles synthesized by FSP may be also applied as a nanomaterial for cancer treatment. For example, Gschwend et al. (2019) used FSP to obtain TiO_2 coated by SiO_2 nanoparticles, which further were applied as a substrate for synthesis of titanium nitride (TiN) nanoparticles coated by SiO_2. Titanium nitride is known photothermal material, which could be applied for photothermal therapy of cancer. Photothermal treatment of cancer is promising alternative to conventional chemotherapy, which is targeted to affect the cancerous cells without destruction of healthy tissues. The procedure is based on the introduction of photothermal agents in the form of nanoparticles to the targeted diseased areas and to induce the temperature increase by the light absorption. The authors claimed that application of $50\,\mu g/mL$ of SiO_2-coated titanium nitride nanoparticles activated by a laser beam at weave length $785\,nm$ was able to destroy cancer cells (Gschwend et al. 2019).

The type of precursors and solvent used for nanoparticles synthesis for medical applications are summarized in the Table 3.7.

ACRONYMS

BET – Brunauer, Emmett, Teller method
FP – Flash Point
FSD – Flame Spray Drying
FSP – Flame Spray Pyrolysis
HEPA – High-Efficiency Particulate Air
LPG – Liquefied Petroleum Gas
PC – Pulse Combustor
PCD – Pulse Combustion Drying
PSD – Particle Size Distribution
PTFE – Polytetrafluoroethylene
SD – Spray Drying
SMD – Sauter Mean Diameter
SSA – Specific Surface Area
TWA – Time-Weighted Average
UV – Ultra-Violet
VOC – Volatile Organic Compounds

TABLE 3.7

Application of FSP for Medical Materials Fabrication

Product	Precursor	Application	Process Parameters	Product Characteristic	Reference
CaP	nd	Drug delivery nanocarrier	Precursor flow rate: 8 and 3 mL/ min, Atomizing O_2 flow rate: 3 and 8 L/min	SSA: 73 and 246 m^2/g, Particles diameter (BET): 26 and 8 nm	Tsikourkitoudi et al. (2020)
CaP	Calcium acetate Solvent: propionic acid Ca/P ratio range: 1.27, 1.38, 1.61, 1.95 and 2.29	Bone substitute and bone- filling cement	nd	SSA (BET) in the range from ca. 40–50 m^2/g, Average particle size: 23 nm	Ataol et al. (2015)
Si-doped WO_3	Ammonium metatungstate hydrate, hexamethyldisiloxane, Solvents: ethanol, diethylene glycol monobutyl ether	Breath acetone sensor for regulation of ketosis during ketogenic diet	Precursor flow rate: 5 mL/min, Atomizing O_2 flow rate: 5 L/min Pilot flame: CH_4 (1.25 L/min), O_2 (3.2 L/min), Additional O_2 flow rate: 5 L/min	Average particle size (BET): 12 nm	Güntner et al. (2018)
SiO_2-coated TiO_2	Titanium, isopropoxide, 2-ethylhexanoic acid and acetonitrile	Material for TiN particles applied for photothermal therapy of cancer	Precursor flow rate: 8 mL/min, Atomizing O_2 flow rate: 3 L/min, Pilot flame: CH_4 (1.5 L/min), O_2 (3.2 L/min)	SSA: from 39.5 to 43.1 m^2/g	Gschwend et al. (2019)

REFERENCES

Abram, C., M. Mezhericher, F. Beyrau, H.A. Stone, and Y. Ju. 2019. "Flame Synthesis of Nanophosphors Using Sub-Micron Aerosols." *Proceedings of the Combustion Institute* 37 (1): 1231–1239. doi:10.1016/j.proci.2018.06.040.

Ajmal, A., I. Majeed, R.N. Malik, H. Idriss, and M.A. Nadeem. 2014. "Principles and Mechanisms of Photocatalytic Dye Degradation on TiO_2 Based Photocatalysts: A Comparative Overview." *RSC Advances* 4 (70): 37003–37026. doi:10.1039/c4ra06658h.

Ataol, S., A. Tezcaner, O. Duygulu, D. Keskin, and N.E. Machin. 2015. "Synthesis and Characterization of Nano-Sized Calcium Phosphates by Flame Spray Pyrolysis, and Their Effect on Osteogenic Differentiation of Stem Cells." *Journal of Nanoparticle Research* 17 (95): 1–14. doi:10.1007/s11051-015-2901-0.

Bahadori, E., M. Rapf, A. Di Michele, and I. Rossetti. 2020. "Photochemical vs. Photocatalytic Azo-Dye Removal in a Pilot Free-Surface Reactor: Is the Catalyst Effective?" *Separation and Purification Technology* 237: 1–11. doi:10.1016/j. seppur.2019.116320.

Beh, G.K., Ch.T. Wang, K. Kim, J. Qu, J. Cairney, Y.H. Ng, A. An, R. Ryoo, A. Urakawa, and W.Y. Teoh. 2020. "Flame-Made Amorphous Solid Acids with Tunable Acidity for the Aqueous Conversion of Glucose to Levulinic Acid." *Green Chemistry* 22 (3): 688–698. doi:10.1039/c9gc02567g.

Birke, W., H.U. Eltz, and F. Schön. 1972. "Verfahren Und Vorrichtung Zum Impragnieren Und Trocken von Textilmaterial." German Patent DE 2214714.

Bunpang, K., A. Wisitsoraat, A. Tuantranont, S. Phanichphant, and Ch. Liewhiran. 2019. "Effects of Reduced Graphene Oxide Loading on Gas-Sensing Characteristics of Flame-Made Bi2WO6 Nanoparticles." *Applied Surface Science* 496: 1–11. doi:10.1016/j. apsusc.2019.143613.

Buss, L., D. Noriler, and U. Fritsching. 2020. "Impact of Reaction Chamber Geometry on the Particle-Residence-Time in Flame Spray Process." *Flow, Turbulence and Combustion* 105: 1055–1086. doi:10.1007/s10494-020-00187-1.

Chomkitichai, W., H. Ninsonti, A. Baba, S. Phanichphant, K. Shinbo, K. Kato, and F. Kaneko. 2014. "Multiple Plasmonic Effect on Photocurrent Generation of Metal-Loaded Titanium Dioxide Composite/Dye Films on Gold Grating Surface." *Surface and Interface Analysis* 46 (9): 607–612. doi:10.1002/sia.5577.

Dahl, P.I., M.S. Thomassen, L.C. Colmenares, A.O. Barnett, S. Lomas, P.E. Vullum, S.M. Hanetho, and T. Mokkelbost. 2015. "Flame Spray Pyrolysis of Electrode Materials for Energy Applications." *MRS Online Proceedings Library (OPL)* 1747: 25–30. doi:10.1557/opl.2015.340.

Daiyan, R., E.C. Lovell, B. Huang, M. Zubair, J. Leverett, Q. Zhang, S. Lim, et al. 2020. "Uncovering Atomic-Scale Stability and Reactivity in Engineered Zinc Oxide Electrocatalysts for Controllable Syngas Production." *Advanced Energy Materials* 2001381: 1–9. doi:10.1002/aenm.202001381.

Ekonek. 2021. "Pulse combustion drying (PCD)." Accessed March 11. https://www.ekonek. eu/en/solutions/pulse-combustion-drying.

Eltz, H.U., G. Petersohn, and F. Schön. 1981. "Alkohole Als Energieträger in Der Textilveredlung : Das ® Remaflam-Verfahren." *Internationales Textil-Bulletin. Weltausgabe Färberei, Druckerei, Ausrüstung* 27 (2): 101–167.

Fujishima, A., and K. Honda. 1972. "Electrochemical Photolysis of Water at a Semiconductor Electrode." *Nature* 238: 37–38. doi:10.1038/238037a0.

Fujiwara, K., Sh. Tada, T. Honma, H. Sasaki, M. Nishijima, and R. Kikuchi. 2019. "Influences of Particle Size and Crystallinity of Highly Loaded CuO/ZrO2 on CO2 Hydrogenation to Methanol." *AIChE Journal* 65: 1–10. doi:10.1002/aic.16717.

Giannouri, M., Th. Kalampaliki, N. Todorova, T. Giannakopoulou, N. Boukos, D. Petrakis, T. Vaimakis, and C. Trapalis. 2013. "One-Step Synthesis of TiO2/Perlite Composites by Flame Spray Pyrolysis and Their Photocatalytic Behavior." *International Journal of Photoenergy* 2013: 729460. doi:10.1155/2013/729460.

Gonchikzhapov, M., and T. Kasper. 2020. "Decomposition Reactions of Fe(CO)5, Fe(C5H5)2, and TTIP as Precursors for the Spray-Flame Synthesis of Nanoparticles in Partial Spray Evaporation at Low Temperatures." *Industrial and Engineering Chemistry Research* 59 (18): 8551–8561. doi:10.1021/acs.iecr.9b06667.

Gröhn, A.J., S.E. Pratsinis, A. Sánchez-Ferrer, R. Mezzenga, and K. Wegner. 2014. "Scale-up of Nanoparticle Synthesis by Flame Spray Pyrolysis: The High-Temperature Particle Residence Time." *Industrial and Engineering Chemistry Research* 53 (26): 10734–10742. doi:10.1021/ie501709s.

Gschwend, P.M., S. Conti, A. Kaech, C. Maake, and S.E. Pratsinis. 2019. "Silica-Coated TiN Particles for Killing Cancer Cells." *ACS Applied Materials & Interfaces* 11: 22550–22560. doi:10.1021/acsami.9b07239.

Guan, Y., G. Deng, Y. Cheng, Zh. Wang, Z. Li, K. Yang, and Y. Yang. 2020. "Facile Preparation of Pd-SiO2 Catalyst through Flame Spray Pyrolysis Method for Enhanced Oxidation of Aromatic Hydrocarbons." *Chemical Physics Letters* 754: 137508. doi:10.1016/j.cplett.2020.137508.

Guild, C., S. Biswas, Y. Meng, T. Jafari, A.M. Gaffney, and S.L. Suib. 2014. "Perspectives of Spray Pyrolysis for Facile Synthesis of Catalysts and Thin Films : An Introduction and Summary of Recent Directions." *Catalysis Today* 238: 87–94. doi:10.1016/j.cattod.2014.03.056.

Güntner, A.T., J.F. Kompalla, H. Landis, S.J. Theodore, B. Geidl, N.A. Sievi, M. Kohler, S.E. Pratsinis, and P.A. Gerber. 2018. "Guiding Ketogenic Diet with Breath Acetone Sensors Guiding Ketogenic Diet with Breath Acetone Sensors." *Sensors* 18: 3655. doi:10.3390/s18113655.

Hembram, K., D. Sivaprakasam, T.N. Rao, and K. Wegner. 2013. "Large-Scale Manufacture of ZnO Nanorods by Flame Spray Pyrolysis." *Journal of Nanoparticle Research* 15 (1461): 1–11. doi:10.1007/s11051-013-1461-4.

Hoverath, A. 1981. Dyeing and Drying with the Remaflam Process." *Textile Chemist & Colorist* 13 (2): 26. http://connection.ebscohost.com/c/articles/31766934/dyeing-drying-remaflam-process.

Huang, Y., S.S.H. Ho, R. Niu, L. Xu, Y. Lu, J. Cao, and Sh. Lee. 2016. "Removal of Indoor Volatile Organic Compounds via Photocatalytic Oxidation: A Short Review and Prospect." *Molecules* 21 (56): 1–20. doi:10.3390/molecules21010056.

Jansanthea, P., W. Chomkitichai, J. Ketwaraporn, P. Pookmanee, and S. Phanichphant. 2019. "Flame Spray Pyrolysis Synthesized Gold-Loaded Titanium Dioxide Photocatalyst for Degradation of Rhodamine B." *Journal of the Australian Ceramic Society* 55 (3): 719–727. doi:10.1007/s41779-018-0283-3.

Joni, I.M., A. Purwanto, F. Iskandar, M. Hazata, and K. Okuyama. 2009. "Intense UV-Light Absorption of ZnO Nanoparticles Prepared Using a Pulse Combustion-Spray Pyrolysis Method." *Chemical Engineering Journal* 155 (1–2): 433–441. doi:10.1016/j.cej.2009.07.011.

Kaewsiri, D., K. Inyawilert, A. Wisitsoraat, A. Tuantranont, S. Phanichphant, and Ch. Liewhiran. 2020. "Single-Nozzle Flame Synthesis of Spinel ZnSnO Nanoparticles for Selective Detection of Formic Acid." *IEEE Sensors Journal* 20 (12): 6256–6262. doi:10.1109/JSEN.2020.2977392.

Kammler, H.K., and S.E. Pratsinis. 1999. "Scaling-up the Production of Nanosized SiO2-Particles in a Double Diffusion Flame Aerosol Reactor." *Journal of Nanoparticle Research* 1 (4): 467–477. doi:10.1023/A:1010080004637.

Kemmler, J.A., S. Pokhrel, L. Mädler, U. Weimar, and N. Barsan. 2013. "Flame Spray Pyrolysis for Sensing at the Nanoscale." *Nanotechnology* 24 (442001): 1–14. doi:10.1088/0957-4484/24/44/442001.

Khamfoo, K., K. Inyawilert, A. Wisitsoraat, A. Tuantranont, S. Phanichphant, and C. Liewhiran. 2020. "Formaldehyde Sensor Based on FSP-Made AgOx-Doped SnO2 Nanoparticulate Sensing Films." *Sensors and Actuators, B: Chemical* 309 (127705): 1–19. doi:10.1016/j.snb.2020.127705.

Kim, J.H., J.K. Kim, and Y.C. Kang. 2020. "Sodium-Ion Storage Performances of MoS2 Nanocrystals Coated with N-Doped Carbon Synthesized by Flame Spray Pyrolysis." *Applied Surface Science* 523 (146470): 1–9. doi:10.1016/j.apsusc.2020.146470.

Kudra, T. 2008. "Pulse-Combustion Drying: Status and Potentials." *Drying Technology* 26 (12): 1409–1420. doi:10.1080/07373930802458812.

Laine, R.M., R. Baranwal, T. Hinklin, D. Treadwell, A. Sutorik, C. Bickmore, K. Waldner, and S.S. Neo. 1999. "Making Nanosized Oxide Powders from Precursors by Flame Spray Pyrolysis." *Key Engineering Materials* 159–160: 17–24. doi:10.4028/www.scientific.net/kem.159-160.17.

Lee, J.H. 2018. "Technological Realization of Semiconducting Metal Oxide-Based Gas Sensors." In *Gas Sensors Based on Conducting Metal Oxides: Basic Understanding, Technology and Applications*, edited by N. Barsa and K. Schierbaum, 167–216. New York: Elsevier. doi:10.1016/B978-0-12-811224-3.00004-4.

Li, H., S. Pokhrel, M. Schowalter, A. Rosenauer, J. Kiefer, and L. Mädler. 2020. "The Gas-Phase Formation of Tin Dioxide Nanoparticles in Single Droplet Combustion and Flame Spray Pyrolysis." *Combustion and Flame* 215: 389–400. doi:10.1016/j.combustflame.2020.02.004.

Liu, Ch., S. Pokhrel, Ch. Tessarek, H. Li, M. Schowalter, A. Rosenauer, M. Eickhoff, Sh. Li, and L. Mädler. 2020. "Rare-Earth-Doped Y4Al2O9 Nanoparticles for Stable Light-Converting Phosphors." *ACS Applied Nano Materials* 3 (1): 699–710. doi:10.1021/acsanm.9b02231.

Liu, Ch., Sh. Li, Y. Zong, Q. Yao, and S.D. Tse. 2017. "Laser-Based Investigation of the Transition from Droplets to Nanoparticles in Flame-Assisted Spray Synthesis of Functional Nanoparticles." *Proceedings of the Combustion Institute* 36: 1109-1117. doi:10.1016/j.proci.2016.06.166.

Liu, S., Ch. Zhou, Y. Wang, E. Yi, W. Wang, J. Kieffer, and R.M. Laine. 2020. "Processing Combustion Synthesized Mg0.5Zr2(PO4)3 Nanopowders to Thin Films as Potential Solid Electrolytes." *Electrochemistry Communications* 116 (106753): 1–6. doi:10.1016/j.elecom.2020.106753.

Liu, X.D., C.W. Cao, and Z.H. Lang. 2001. "Heat Transfer between Materials and Unsteady Airflow from a Helmholtz Type Combustor." *Drying Technology* 19 (8): 1939–1948.

Mädler, L., and S.E. Pratsinis. 2002. "Bismuth Oxide Nanoparticles by Flame Spray Pyrolysis." *Journal of the American Ceramic Society* 85 (7): 1713–1718. doi:10.1111/j.1151-2916.2002.tb00340.x.

Mädler, L., H.K. Kammler, R. Mueller, and S.E. Pratsinis. 2002. "Controlled Synthesis of Nanostructured Particles by Flame Spray Pyrolysis." *Journal of Aerosol Science* 33 (2): 369–389. doi:10.1016/S0021-8502(01)00159-8.

Mädler, L., W.J. Stark, and S.E. Pratsinis. 2002. "Flame-Made Ceria Nanoparticles." *Journal of Materials Research* 17 (6): 1356–1362. doi:10.1557/JMR.2002.0202.

Meng, L., and H. Zhao. 2020. "Low-Temperature Complete Removal of Toluene over Highly Active Nanoparticles CuO-TiO2 Synthesized via Flame Spray Pyrolysis." *Applied Catalysis B: Environmental* 264 (118427): 1–10. doi:10.1016/j.apcatb.2019.118427.

Meng, X., W. De Jong, and T. Kudra. 2016. "A State-of-the-Art Review of Pulse Combustion: Principles, Modeling, Applications and R&D Issues." *Renewable and Sustainable Energy Reviews* 55: 73–114. doi:10.1016/j.rser.2015.10.110.

Mueller, R., L. Mädler, and S.E. Pratsinis. 2003. "Nanoparticle Synthesis at High Production Rates by Flame Spray Pyrolysis." *Chemical Engineering Science* 58 (10): 1969–1976. doi:10.1016/S0009-2509(03)00022-8.

Mujumdar, A.S. 2004. "Research and Development in Drying: Recent Trends and Future Prospects." *Drying Technology* 22 (1–2): 1–26. doi:10.1081/DRT-120028201.

Oguzlar, S. 2020. "Development of Highly Sensitive [Ru(Bpy)3]2+ - Based Optical Oxygen Sensing Thin Films in the Presence with Fe3O4 and Fe3O4@Ag NPs." *Optical Materials* 101 (109772): 1–7. doi:10.1016/j.optmat.2020.109772.

Ohkawara Kakohki Co., Ltd. 2020. "OC - Drying Setups." Accessed May 26. https://www.oc-sd.co.jp/english/product/?url=product.html#products1.

Pineau, N.J., S.D. Keller, A.T. Güntner, and S.E. Pratsinis. 2020. "Palladium Embedded in SnO2 Enhances the Sensitivity of Flame-Made Chemoresistive Gas Sensors." *Microchimica Acta* 187: 96. doi:10.1007/s00604-019-4080-7.

Pratsinis, E. 1998. "Flame Aerosol Synthesis of Ceramic Powders." *Progress in Energy and Combustion Science* 24 (3): 197–219. doi:10.1016/s0140-6701(98)94112-1.

Psathas, P., Y. Georgiou, C. Moularas, G.S. Armatas, and Y. Deligiannakis. 2020. "Controlled-Phase Synthesis of Bi2Fe4O9 & BiFeO3 by Flame Spray Pyrolysis and Their Evaluation as Non-Noble Metal Catalysts for Efficient Reduction of 4-Nitrophenol." *Powder Technology* 368: 268–277. doi:10.1016/j.powtec.2020.04.059.

Rehkopf, J., and D. Mirko. 2017. "Pulse Combustion Drying of Proteins." US Patent US9809619B2.

Robin, J.-P., and P. Lenoir. 2009. "Flame Dryer." US Patent US2009/0007453 A1.

Sakar, N., H. Gergeroglu, S.A. Akalin, S. Oguzlar, and S. Yildirim. 2020. "Synthesis, Structural and Optical Characterization of Nd: YAG Powders via Flame Spray Pyrolysis." *Optical Materials* 103 (109819): 1–10. doi:10.1016/j.optmat.2020.109819.

San Martin, D., S. Ramos, and J. Zufía. 2016. "Valorisation of Food Waste to Produce New Raw Materials for Animal Feed." *Food Chemistry* 198: 68–74. doi:10.1016/j.foodchem.2015.11.035.

Sarkar, Mi.R., N. Wachter, P. Patka, and L. Kinzl. 2001. "First Histological Observations on the Incorporation of a Novel Calcium Phosphate Bone Substitute Material in Human Cancellous Bone." *Journal of Biomedical Materials Research* 58 (3): 329–334. doi:10.1002/1097-4636(2001)58:3<329::AID-JBM1025>3.0.CO;2-9.

Stark, W.J., L. Madler, and S.E. Pratsinis. 2007. "Flame Made Methal Oxides." US Patent US7211236 B2.

Strumillo, Cz., I. Zbiciński, I. Smucerowicz, and C. Crowe. 1999. "An Analysis of a Pulse Combustion Drying System." *Chemical Engineering and Processing: Process Intensification* 38 (4–6): 593–600. doi:10.1016/S0255-2701(99)00060-4.

Sukee, A., A.A. Alharbi, A. Staerz, A. Wisitsoraat, C. Liewhiran, U. Weimar, and N. Barsan. 2020. "Effect of AgO Loading on Flame-Made LaFeO3 p-Type Semiconductor Nanoparticles to Acetylene Sensing." *Sensors and Actuators, B: Chemical* 312 (127990): 1–11. doi:10.1016/j.snb.2020.127990.

Teoh, W.Y., R. Amal, and L. Mädler. 2010. "Flame Spray Pyrolysis: An Enabling Technology for Nanoparticles Design and Fabrication." *Nanoscale* 2 (8): 1324–1347. doi:10.1039/c0nr00017e.

Tsikourkitoudi, V., J. Karlsson, P. Merkl, E. Loh, B. Henriques-Normark, and G.A. Sotiriou. 2020. "Flame-Made Calcium Phosphate Nanoparticles with High Drug Loading for Delivery of Biologics." *Molecules* 25 (1747): 1–17. doi:10.3390/molecules25071747.

Ulrich, G.D. 1984. "Flame Synthesis of Fine Particles." *Chemical & Engineering News* 62 (32): 22–29.

Wang, L., F.D. Cui, and H. Sunada. 2007. "Improvement of the Dissolution Rate of Nitrendipine Using a New Pulse Combustion Drying Method." *Chemical and Pharmaceutical Bulletin* 55 (8): 1119–1125. doi:10.1016/S1773-2247(09)50018-4.

Wawrzyniak, P., I. Zbicinski, and M. Sobulska. 2017. "Applications: Drying of Materials." In *CRC Handbook of Thermal Engineering*, 2nd ed., 1306–1337. Boca Raton, FL: CRC Press. doi:10.4324/9781315119717.

Wei, J., Sh. Li, Y. Ren, Y. Zhang, and S.D. Tse. 2019. "Investigating the Role of Solvent Formulations in Temperature-Controlled Liquid-Fed Aerosol Flame Synthesis of YAG-Based Nanoparticles." *Proceedings of the Combustion Institute* 37: 1193–1201. doi:10.1016/j.proci.2018.07.068.

Weyell, P., H.D. Kurland, T. Hülser, J. Grabow, F.A. Müller, and D. Kralisch. 2020. "Risk and Life Cycle Assessment of Nanoparticles for Medical Applications Prepared Using Safe- And Benign-by-Design Gas-Phase Syntheses." *Green Chemistry* 22 (3): 814–827. doi:10.1039/c9gc02436k.

Wu, Zh., L. Yue, Zh. Li, J. Li, A.S. Mujumdar, and J.A. Rehkopf. 2014. "Pulse Combustion Spray Drying of Egg White: Energy Efficiency and Product Quality." *Food and Bioprocess Technology* 8 (1): 148–157. doi:10.1007/s11947-014-1384-9.

Xiao, Zh., X. Xie, Y. Yuan, and X. Liu. 2008. "Influence of Atomizing Parameters on Droplet Properties in a Pulse Combustion Spray Dryer." *Drying Technology* 26 (4): 427–432. doi:10.1080/07373930801929235.

Xiong, Zh., Z. Lei, Z. Xu, X. Chen, B. Gong, Y. Zhao, H. Zhao, J. Zhang, and Ch. Zheng. 2017. "Flame Spray Pyrolysis Synthesized ZnO/CeO2 Nanocomposites for Enhanced CO2 Photocatalytic Reduction under UV-Vis Light Irradiation." *Journal of CO$_2$ Utilization* 18 (March): 53–61. doi:10.1016/j.jcou.2017.01.013.

Xiong, Zh., Z. Xu, Y. Li, L. Dong, J. Wang, J. Zhao, X. Chen, Y. Zhao, H. Zhao, and J. Zhang. 2020. "Incorporating Highly Dispersed and Stable Cu+ into TiO2 Lattice for Enhanced Photocatalytic CO2 Reduction with Water." *Applied Surface Science* 507 (145095): 1–8. doi:10.1016/j.apsusc.2019.145095.

Xu, L., S.M. Li, and H. Sunada. 2007. "Preparation and Evaluation of Ibuprofen Solid Dispersion Systems with Kollidon Particles Using a Pulse Combustion Dryer System." *Chemical and Pharmaceutical Bulletin* 55 (11): 1545–1550. doi:10.1248/cpb.55.1545.

Yudha, C.S., S.U. Muzayanha, M. Rahmawati, H. Widiyandari, W. Sutopo, M. Nizam, S.P. Santosa, and A. Purwanto. 2020. "Fast Production of High Performance LiNi0.815Co0.15Al0.035O2 Cathode Material via Urea-Assisted Flame Spray Pyrolysis." *Energies* 13 (2757): 1–17.

Zbicinski, I. 2002. "Equipment, Technology, Perspectives and Modeling of Pulse Combustion Drying." *Chemical Engineering Journal* 86 (1–2): 33–46. doi:10.1016/S1385-8947(01)00269-8.

Zbicinski, I., M. Benali, and T. Kudra. 2002. "Pulse Combustion: An Advanced Technology for Efficient Drying." *Chemical Engineering and Technology* 25 (7): 687–691. doi:10.1002/1521-4125(20020709)25:7<687::aid-ceat687>3.0.co;2-%23.

Zbicinski, I., T. Kudra, and X. Liu. 2014. "Pulse Combustion Drying." In *Modern Drying Technology*, edited by E. Tsotsas and A.S. Mujumdar, Vol. 5, 27–56. Weinheim, Germany: Wiley-VCH Verlag GmbH & Co. KGaA. doi:10.1002/9783527631704.ch02.

Zhang, J., S. Xu, Kh.I. Hamad, A.M. Jasim, and Y. Xing. 2020. "High Retention Rate NCA Cathode Powders from Spray Drying and Flame Assisted Spray Pyrolysis Using Glycerol as the Solvent." *Powder Technology* 363: 1–6. doi:10.1016/j.powtec.2019.12.057.

Zhonghua, W., and A.S. Mujumdar. 2006. "R&D Needs and Opportunities in Pulse Combustion and Pulse Combustion Drying." *Drying Technology* 24 (11): 1521–1523. doi:10.1080/07373930600961520.

Zhonghua, W., W. Long, L. Zhanyong, and A.S. Mujumdar. 2012. "Atomization and Drying Characteristics of Sewage Sludge inside a Helmholtz Pulse Combustor." *Drying Technology* 30 (10): 1105–1112. doi:10.1080/07373937.2012.683122.

4 Safety, Energy, Environmental Issues, and Perspectives of FSD Technique Development

4.1 SAFETY OPERATION DURING FSD

Flame spray drying (FSD) is a complex process where spray drying is combined with liquid fuel combustion; therefore, the safety aspects of process should include comprehensive analysis and identification of hazards driven by both drying and combustion processes and elaboration of required protection measures.

Conventional drying processes are highly predisposed to fire and explosion hazards due to the following reasons:

- External heat sources are utilized for water evaporation
- Many of the dried materials are combustible and prone to dust explosions when contacting with the air at elevated temperatures
- Presence of oxygen if air is applied as a drying medium
- Process is performed in confined space – drying chamber (Markowski and Mujumdar 2006).

Another source of fire and explosion hazards is development of wall deposits, which when overheated increase the risk of ignition of powder accumulated at the dryer walls.

In the FSD, additional source of fire and explosion hazards is the presence of ignition source (direct fire) and flammable liquids and gases within the drying chamber.

In FSD process, the following safety measures should be applied:

- The evaluation of material characteristic such as explosiveness, combustibility, dust explosion characteristics (lower and upper dust explosivity limits, minimum ignition temperature, and minimum ignition energy of dust cloud) before running the drying process. The minimum ignition energy for most air/dust mixtures is about 100 mJ (Andrews and Phylaktou 2010), for example, 30 mJ for sugar and 120 mJ for cocoa (Markowski and Mujumdar 2006). The minimum ignition temperature characterizes the temperature value, which should not be exceeded during the dryer operation, the typical values of minimum ignition

DOI 10.1201/9781003100386-4

temperatures are as follows: 370°C for sugar and 500°C for cocoa (Markowski and Mujumdar 2006).

- Selection of proper process parameters to ensure safe operation. Air temperature within the drying chamber as well as within the air/powder separation units should be kept below the minimal ignition temperature of material being dried. One of the most common explosion prevention methods is application of ventilation and dust concentration dilution by excess amount of air to shift air/dust mixture ratio from the flammable range. During FSD, the high air flow rate is applied (from about 150 to 550 Nm^3/h) to achieve equivalence ratio (stoichiometric air/fuel ratio to the current air/fuel ratio) in the range from ca. 0.03 to 0.3. Additionally, to avoid any leakages of dust into the working area, it is recommended to operate the spray dryer under the slight underpressure; during FSD, the pressure drop within the drying chamber should be in the range from 150 to about 500 Pa.
- Proper dryer design. To avoid the ignition of powder wall deposits, it is recommended to apply appropriate dryer diameter, which could ensure sufficient distance between flame and dryer walls. The flame spray dryer as well as dust collection and separation equipment should be located outside of the building at the open space or within the fire-resistant enclosure (Markowski and Mujumdar 2006).
- Application of proper fire and explosion-protection measures. Since in FSD process, due to nature of the process, the ignition source and flammable vapors cannot be eliminated from the drying chamber, the appropriate explosion-protection measures should be installed. Prevention measures, such as application of inert gases as a drying medium, are limited since the oxygen should be provided to the drying chamber for spray combustion. Therefore, it is crucial to install appropriate control and protection systems as well as develop and implement the system of alarms, faults, and stop sequences during the operation of flame spray dryer. The typical explosion-protection measures include two options: explosion suppression and explosion venting (Andrews and Phylaktou 2010). Explosion suppression measure includes application of dry powder fire extinguisher or high rate discharge (HRD) vessels, the action of which starts if pressure sensor detects the pressure rise within the drying chamber. The sudden pressure rise associated with explosion falls within the pressure range from 50 to 100 mbar (Andrews and Phylaktou 2010). The signal from the pressure sensor sent to the automatically controlled valve to discharge of extinguishing substance (typically sodium bicarbonate or mono-ammonium phosphate) into the drying chamber and elements of dedusting systems such as cyclones or filters or pipes connecting equipment units. Explosion venting measures include installation of vent in the dryer walls, which opens if certain static pressure in the dryer is reached. For proper application of explosion vents, the static pressure at which vent opens should be far below the maximum explosion pressure of

the dryer. For typical dryers, the maximum explosion pressure is below 900 kPa (Markowski and Mujumdar 2006). Another requirement for proper explosion vent application is precise determination of the vent area. Andrews and Phylaktou (2010) provide the guidance for vent design for two types of vessels: compact vessels with length-to-diameter ratio below 2 and long vessels with length-to-diameter ratio above 2. In general, the long vessels such as spray dryers require larger vent area for the same dryer volume (Wawrzyniak et al. 2012; Polanczyk, Wawrzyniak, and Zbicinski 2013).

- Application of flame detectors. The application of control loop, which includes flame detector in coupling with feeding pump motor frequency converter, will ensure that in case if fuel is not ignited in FSD chamber, then the flame detector will send a signal to control system, which will immediately shut down the fuel feeding pump. This stop sequence is required to avoid the situation when the concentration of fuel vapor in the drying chamber will increase drastically and may result in the uncontrolled spread of fire and further explosion.

An example of location of fire and explosion protection units in FSD system is presented in the Figure 4.1. Pressure detector PIC with short reaction time is placed inside the drying chamber. In case of explosion, even small pressure increase of 0.03–0.15 bar will be recognized by pressure detector and the signal will be sent to control unit in 5–35 ms (technical data of *RSBP spol. s r.o.*, Ostrava, Czech Rebublic). The signal sent by pressure detector will force the fast opening of vent on the HRD container unit to allow for supply of the explosion suppressing substance into the drying chamber. Additional HRD barrier is located on the pipe connecting the dryer and cyclone to avoid the spread of fire and explosion propagation to other parts of installation. On the top of the dryer an explosion venting device is located to relieve an excess pressure in case of explosion. The flame detector XC mounted in the drying

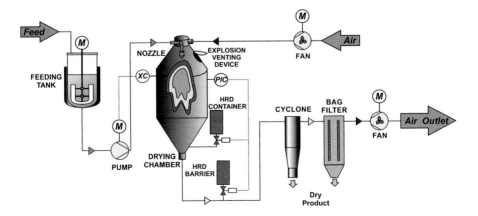

FIGURE 4.1 Location of fire and explosion protection units in FSD system. (Elaborated based on technical data of RSBP spol. s r.o. (Ostrava, Czech Republic), www.rsbp.cz.)

chamber will force the shutdown of the feeding pump if ignition system failed and the flame will not be detected for about 5 s.

4.2   ENERGY CONSUMPTION

Analysis of energy efficiency in FSD process and energy consumption during standard spray drying (SSD) process was carried out by Sobulska (2019) and Piatkowski, Taradaichenko, and Zbicinski (2015). Due to substantial differences in the principles and operation parameters between the SSD and FSD techniques, the process parameters and feed compositions for experimental tests must be carefully selected to provide equivalent conditions for both drying processes.

The authors carried out standard spray drying and flame spray drying tests with addition 35 and 45 wt.% of ethanol in the feed to compare energy consumption in both processes (Table 4.1).

The feed rate for SSD was equal to 5 kg/h, whereas in the flame spray drying tests, feed rates were adjusted to achieve equivalent moisture evaporation rates for both drying techniques.

In the FSD process, heat for moisture evaporation is generated in the atomization zone, so air inlet temperature is not a suitable parameter to compare FSD and SSD process. Therefore, only air temperature at the outlet of the dryer and air flow rates were kept similar during FSD and SSD tests. The temperature of sprayed solutions was 70°C in all experimental tests. Atomization pressure in the experiments varied from 1.1 to 5.2 MPa.

In the literature, various indexes are applied to characterize the drying process in terms of energy aspects, among which energy efficiency and specific energy consumption are the most common. Energy efficiency is defined as a ratio of the energy used for moisture evaporation to the total energy supplied to the dryer (Kudra 2007). Baker and Mckenzie (2005) applied specific energy consumption (energy consumed for evaporation of unit mass of water) to determine the energy performance of industrial spray dryers.

In Sobulska (2019) and Piatkowski, Taradaichenko, and Zbicinski (2015), the specific energy consumption in SSD process, E_{SSD} (MJ/kg$_{H2O}$), and in FSD process, E_{FSD} (MJ/kg$_{H2O}$) was calculated as proposed by Al-Mansour, Al-Busairi, and Baker (2011):

$$E_{SSD} = \frac{Q_{SSD}}{E_v} = \frac{Q_{SSD}}{F_s\left(X_i - X_0\right)} \tag{4.1}$$

$$E_{FSD} = \frac{Q_{FSD}}{E_v} = \frac{Q_{FSD}}{F_s\left(X_i - X_0\right)} \tag{4.2}$$

where Q_{SSD} and Q_{FSD} are the heat inputs to the dryer in SSD and FSD process resp. [MJ/h], E_v is the amount of evaporated moisture, which is calculated from the flow rate of solid F_s and moisture content of the feed X_0 and product X_i.

For convective dryer, heat input Q_{SSD} is equal to power supplied to the heater (Kudra 2012). The power consumed in the heater in SSD process can be determined

TABLE 4.1

Process Parameters for Different Spray Drying Tests – Analysis of Energy Efficiency (Solution A – Solids Content in Fuel-free Solution 30 wt.%, Solution B – 40 wt.%, Solution C – 50 wt.%)

Test	Flow Rate of Water (for Evaporation) (kg/h)	Flow Rate of Solid (for Drying from the Solution) (kg/h)	Feed Composition (wt.%)			Feed Rate (kg/h)	Air Flow Rate (N m³/h)	Air Outlet Temperature (°C)
			Ethanol	Maltodextrin	Water			
			Solution A					
1	3.5	1.5	0[a]	30	70	5.0	455	94
2			45	16.5	38.5	9.1	455	94
			Solution B					
3	3.0	2.0	0	40	60	5.0	390	83
4			35	26	39	7.7	390	83
5			0	40	60	5.0	550	103
6			45	22	33	9.1	550	103
			Solution C					
7	2.5	2.5	0	50	50	12.0	460	132
8			35	32.5	32.5	7.7	460	132

Source: Based on Piatkowski, Taradaichenko, and Zbicinski (2015).

[a] Standard spray drying.

from the electrical current of individual heating elements and voltage of the electric system.

$$Q_{SSD} = P * 3,600 = I * U * 3,600 \tag{4.3}$$

where P is electric power input, I is electrical current, U is voltage of the system.

The energy consumption during FSD can be calculated from net enthalpy of combustion of fuel and fuel flow rate assuming that flammable component in the sprayed stream has been evaporated and burned completely (equation 4.5). The net enthalpy of combustion of fuel may be calculated from the fuel concentration and net enthalpy of combustion of each component of the fuel (ethanol – 26.85 MJ/kg, 1-propanol – 30.68 MJ/kg, and 2-propanol – 30.45 MJ/kg) (Green and Perry 2008) as provided by equation (4.4).

$$\Delta H_f = Y_{et} \cdot \Delta H_{et} + Y_{prop1} \cdot \Delta H_{prop1} + Y_{prop2} \cdot \Delta H_{prop2} \tag{4.4}$$

$$Q_{FSD} = F_f * \Delta H_f \tag{4.5}$$

Total heat input in SSD process Q_{SSD} includes heat for moisture evaporation Q_{vap}, heat loss in heater η_{heater}, heat loss in drying tower η_{drier}, heat loss with exhaust air η_{air} and dried product $\eta_{product}$, as shown in the formula:

$$Q_{SSD} = Q_{vap} + \eta_{heater} + \eta_{drier} + \eta_{air} + \eta_{product} \tag{4.6}$$

In FSD process, heat is generated in drying tower; therefore, total heat input in FSD process Q_{FSD} includes heat for moisture evaporation Q_{vap}, heat loss in drying tower η_{drier}, heat loss with exhaust air η_{air} and dried product $\eta_{product}$, as shown in the equation 4.7:

$$Q_{FSD} = Q_{vap} + \eta_{drier} + \eta_{air} + \eta_{product} \tag{4.7}$$

To compare the SSD and FSD processes, total heat inputs for both processes must be used to calculate energy consumption during drying process (equations 4.1 and 4.2).

Comparison of the amount of energy consumed for moisture evaporation during FSD with energy consumption during the SSD process carried out at the equivalent operational parameters is shown in Table 4.2 (Piatkowski, Taradaichenko, and Zbicinski 2015). The energy consumption in both processes ranged from 24 to 50 MJ per 1 kg of evaporated moisture, which is higher than that in the industrial dryers. Baker and Mckenzie (2005) reported that for industrial spray dryers, the energy consumption varied from 3 to 20 MJ/kg_{H2O}, whereas for dryers with low throughput (less than 1 t_{H2O}/h), these values can be even four to five times higher.

The authors (Piatkowski, Taradaichenko, and Zbicinski 2015) concluded that in the SSD process, the amount of energy consumed per 1 kg of evaporated moisture is

TABLE 4.2

Comparison of Energy Consumption in the Standard Spray Drying and FSD Processes in Similar Operating Parameters

Standard Spray Drying		Flame Spray Drying	
Test[a]	Specific Energy Consumption (E_{SSD} MJ/kg$_{H2O}$)	Test[a]	Specific Energy Consumption (E_{FSD} MJ/kg$_{H2O}$)
3	36.11	4	23.92
7	38.46	8	28.53
1	33.75	2	30.85
5	50.09	6	35.96

Source: Based on Piatkowski, Taradaichenko, and Zbicinski (2015).

[a] Test numbers corresponds to tests in Table 4.1.

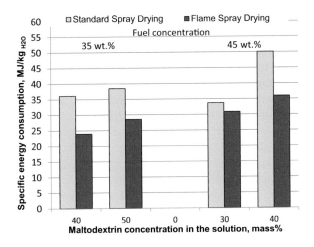

FIGURE 4.2 Energy consumption in the standard spray drying and FSD processes.

higher than that in the FSD process (Table 4.2, Figure 4.2). Depending on the process parameters, the energy consumption of FSD was 5% up to 33% lower than that in the SSD process.

4.3 ENVIRONMENTAL PROTECTION

There are two main sources of pollutions during FSD process: air contamination by dry microparticles and air pollution due to combustion process. Exhaust air after spray drying process should be separated properly from the dry particles in order to minimize product losses and undesired discharge of particles to atmosphere, which cause environmental problems. In the European Union countries, the powder emission for spray drying installation is limited to 10 mg/Nm³ of exhaust air (Pisecky 2012). The air separation equipment typically applied in spray drying includes cyclones, wet scrubbers, and bag filters. Application of cyclones only reduces powder emission to about 200–400 mg/Nm³; therefore, to meet the emission requirements, the cyclones are commonly combined with bag filters or wet scrubbers, which allows to decrease the powder emission level to 5–20 mg/Nm³ (Pisecky 2012).

All combustion processes generate different quantities of air pollutions, such as CO_2 – the main product of combustion reaction as well as nitrogen oxides due to nitrogen/oxygen reaction at elevated temperatures.

CO_2, the main greenhouse gas, is produced during combustion of any fuel via two routes: complete and incomplete fuel combustion. During complete combustion, where fuel is completely oxidized in combustion process, the CO_2, H_2O, and SO_2 reaction products are generated. In case of incomplete combustion, the additional air polluting substances are CO, aldehydes, ketones, soot, and other undesirable products (Bai and Karthik 2010).

The nitrogen oxides emitted in combustion process are referred to as NO_x and present the sum of nitrogen oxide NO and nitrogen dioxide NO_2 emissions. There are two main routes of NO_x formation during combustion of organic fuels: formation from atmospheric N_2 present in air, NO_x formation from nitrogen compounds present in fuel (Konnov et al. 2010). In general, the following factors increase the formation of NO_x during combustion process: high combustion temperatures, high heat transfer rates, excess amount of air, and low residence time in the combustion chamber (Pisecky 2012). Decrease of maximal flame temperature results in lower NO_x emissions (U.S. Department of Energy National Energy Technology 2003).

One of the main advantages of FSD process is possibility to substitute the conventional energy sources such as fossil fuels by biofuels such as bioethanol or vegetable oils, which can be obtained from renewable energy sources. Compared with conventional fossil fuels, bioethanol offers advantage of high octane number – 108, whereas gasoline octane number is in the range from 95 to 98 (Manzetti and Andersen 2015). Moreover, bioethanol has higher latent heat, which increases the volumetric efficiency of combustors (Masum et al. 2013). Replacement of hydrocarbon fuels by biofuels is one of the measures, which could be applied for reduction of CO_2 emissions. The amount of CO_2 generated during combustion process depends on the chemical structure of the fuel burnt, i.e., content of carbon. Thus, the maximum theoretical percentage of CO_2 (mole fraction) of ethanol is 11.01%, which is lower compared with natural gas – 11.8%, oil – 16.5%, coal – 17.0%, wood – 19.1% (Bai and Karthik 2010).

In the literature, contradictory data on NO_x emissions for ethanol combustion are reported (Masum et al. 2013): several studies claimed lower NO_x emissions for ethanol compared with conventional fuels (Rajan 1984; Tavares et al. 2011); however, there are some studies reporting increased NO_x level for ethanol combustion (Hsieh et al. 2002; Najafi et al. 2009). The comparative study of bioethanol and diesel combustion in the industrial gas turbine combustor showed lower NO_x concentrations for bioethanol (Sallevelt et al. 2014).

Moreover, the literature data shows that bioethanol reduces the amounts of particulate, i.e. soot, emission from combustion due to high oxygen content of 35% (Manzetti and Andersen 2015).

Vegetable oils, for example, sunflower oil, rapeseed oil, due to high energy content could also be applied as a fuel coming from renewable energy sources. In general, direct application of vegetable oils as a fuel reduces NO_x pollutions, however, might increase CO and HC emissions (Altin, Çetinkaya, and Yücesu 2001). The combustion emissions could be reduced if vegetable oils are processed into biodiesel, i.e., oil methyl esters. Thus, Shirneshan (2013) reported that biodiesel produced from fry waste vegetable oil produced lower emissions of hydrocarbon (HC) and CO and increased emissions of CO_2 and NO_x compared with conventional diesel fuel.

4.4 ADVANTAGES AND DISADVANTAGES, PERSPECTIVES, AND FURTHER DEVELOPMENT OF FLAME SPRAY DRYING TECHNIQUE

Summing up, we may point out the following advantages of a novel FSD technique over conventional spray drying:

- FSD allows to reduce energy consumption from 5% to 33% compared with SSD method; moreover, the energy performance of FSD process still could be improved by optimalization of process parameters and decrease of fuel concentration in the feed.
- FSD requires lower investment costs compared with SSD as application of auxiliary equipment such as air heaters is not necessary. Moreover, FSD process could be operated independently from the conventional energy supply system, which might be advantageous for countries with high production rate of biofuels: bioethanol and biodiesel.
- FSD offers possibility of direct application of renewable energy sources such as biofuels without significant and costly modification of spray drying plant.
- FSD is recommended for thermally resistant products such as ceramic powder or metal oxides; nevertheless, FSD could be successfully applied for food products such as maltodextrin or coffee, if process parameters are properly selected and controlled.
- FSD can be an alternative to spray drying of feedstock containing flammable solvents, which is currently carried out in the closed-loop systems.
- FSD due to specific flow hydrodynamics generated in the combustion chamber, i.e., oscillating recirculation zones, allows for production of agglomerated product in concurrent spray drying process configuration. Powder agglomeration is limited in standard co-current spray drying process and requires either application of multiple nozzles or multistage drying, i.e., spray drying combined with agglomeration stage.
- Addition of low-viscosity flammable component such as ethanol to the feed during FSD improves atomization properties for high viscosity and high solid content feed solutions.

The main disadvantages of FSD method are as follows:

- High requirement for safety process operation, application of explosion-protection measures: flame detectors, accurate pressure sensors, fire and explosion-protection measures.
- High requirements for thermally resistant materials for construction of drying chamber and atomization nozzles due to extreme operation parameters: high temperatures, high frequency of temperature oscillations.
- Problem to handle feedstock with high solid content, in case of thermally sensitive products thermal degradation of dry material increases with increase of solid content in the feed.

The research work is needed for further optimization of FSD method in terms of flame stabilization, energy consumption, and improvement of final product properties.

Further experimental work on FSD process could be focused on the combustion and pollutants formation mechanism via application of advanced measuring techniques for determination of species concentration within the drying chamber. In FSD the concentration of fuel vapor, oxygen, CO_2, H_2O, as well as main polluting substances such as CO, NO_x, HC could be analyzed by such techniques as coherent anti-Stokes Raman scattering, Raman spectroscopy, and laser-induced fluorescence, etc.

Standard CFD models can be used for scaling-up of FSD. The accuracy of CFD models of FSD process could be improved in terms of the following:

- Precise description of reaction kinetics of fuel combustion including multiple reaction steps with intermediate products and species, instead of assumption of single one-step reaction of combustion.
- Accurate model of droplet evaporation covering first drying period: evaporation of multicomponent droplet (water – ethanol, nonideal mixture), second drying stage: locking point – crust formation resulting in decrease of drying rate.
- Implementation of particle structure formation model via agglomeration and puffing due to presence of particles recirculation zones and rapid droplet drying at elevated temperatures.
- Extension of the CFD models to transient state to account for pressure and temperature fluctuation during FSD process.
- Coupling of CFD model with dried material degradation kinetics model for accurate prediction of final product quality.

Another route of improvement of the FSD process is to utilize the idea of intermittent drying process for further reduction of energy consumption required for dewatering process. In the intermittent drying, the supply of thermal energy is controlled either by varying the drying process parameters (airflow rate, air temperature, humidity, or operating pressure) or by changing the mode of heat supply (convection, radiation, microwave) (Kumar, Karim, and Joardder 2014). Application of intermittent drying under periodic drying conditions allows for significant reduction of the energy consumption compared with stationary conditions (Kowalski and Pawłowski 2011). Thus, as it has been described in the Chapter 3, pulse combustion drying, which uses periodic heat input generated by intermittent combustion of solid, liquid, and gaseous fuels, has found applications in the industry due to significantly reduced energy demand (e.g., Ekonek).

Therefore, further development of FSD techniques may be focused on the application of dewatering of dispersed systems in the flame of pulsed nature, for example, introducing the pulsed flow of flammable component into the drying chamber with subsequent ignition and combustion of the sprayed stream.

ACRONYMS

FSD – Flame Spray Drying
HRD – High Rate Discharge
SSD – Standard Spray Drying

NOMENCLATURE

E_{FSD} = specific energy consumption in FSD (MJ/kg_{H2O})
E_{SSD} = specific energy consumption in SSD (MJ/kg_{H2O})
E_v = amount of evaporated moisture (kg/h)
F_f = flow rate of fuel (kg/h)
F_s = flow rate of solid (kg/h)
I = electrical current (A)
P = electric power input (W)
Q_{FSD} = heat inputs to the dryer in FSD (MJ/h)
Q_{SSD} = heat inputs to the dryer in SSD (MJ/h)
Q_{vap} = heat for moisture evaporation (MJ/h)
U = voltage of the system (V)
X_0 = moisture content of the feed ($kg_{H2O}/kg_{dry\ material}$)
X_i = moisture content of the product ($kg_{H2O}/kg_{dry\ material}$)
Y = mass fraction
ΔH_{et} = net enthalpy of combustion of ethanol (MJ/kg)
ΔH_f = net enthalpy of combustion of fuel (MJ/kg)
ΔH_{prop1} = net enthalpy of combustion of propanol-1 (MJ/kg)
ΔH_{prop2} = net enthalpy of combustion of propanol-2 (MJ/kg)
η = heat loss (MJ/h)

REFERENCES

Al-Mansour, H.E., B.H. Al-Busairi, and C.G.J Baker. 2011. "Energy Consumption of a Pilot-Scale Spray Dryer." *Drying Technology* 29 (16): 1901–1910. doi:10.108 0/07373937.2011.595563.

Altin, R., S. Çetinkaya, and H.S. Yücesu. 2001. "Potential of Using Vegetable Oil Fuels as Fuel for Diesel Engines." *Energy Conversion and Management* 42 (5): 529–538. doi:10.1016/S0196-8904(00)00080-7.

Andrews, G.E., and H.N. Phylaktou. 2010. "Explosion Safety." In *Handbook of Combustion. Vol.1: Fundamentals and Safety*, edited by M. Lackner, F. Winter, and A.K. Agarwal, 377–413. Weinheim, Germany: Wiley-VCH Verlag GmbH & Co. KGaA.

Bai, H., and M. Karthik. 2010. "CO2 Greenhouse Gas Formation and Capture." In *Handbook of Combustion. Vol. 2:Combustion Diagnostics and Pollutants*, edited by M. Lackner, F. Winter, and A.K. Agarwal, 375–402. Weinheim, Germany: Wiley-VCH Verlag GmbH & Co. KGaA.

Baker, C.G.J, and K.A Mckenzie. 2005. "Energy Consumption of Industrial Spray Dryers." *Drying Technology* 23 (1–2): 365–386. doi:10.1081/DRT–200047665.

Ekonek. 2021. "Pulse combustion drying (PCD)." Accessed March 11. https://www.ekonek.eu/en/solutions/pulse-combustion-drying.

Green, D.W., and R.H. Perry. 2008. *Perry's Chemical Engineers' Handbook*. New York: McGraw-Hill.

Hsieh, W.D., R.H. Chen, T.L. Wu, and T.H. Lin. 2002. "Engine Performance and Pollutant Emission of an SI Engine Using Ethanol–Gasoline Blended Fuels." *Atmospheric Environment* 36: 403–410. doi:10.1515/tjj.2001.18.1.1.

Konnov, A.A., T. Javed, H. Kassman, and N. Irfan. 2010. "NOx Formation, Control and Reduction Techniques." In *Handbook of Combustion. Vol. 2:Combustion Diagnostics and Pollutants*, edited by M. Lackner, F. Winter, and A.K. Agarwal, 439–464. Weinheim, Germany: Wiley-VCH Verlag GmbH & Co. KGaA.

Kowalski, S.J., and A. Pawłowski. 2011. "Energy Consumption and Quality Aspect by Intermittent Drying." *Chemical Engineering and Processing: Process Intensification* 50 (4): 384–390. doi:10.1016/j.cep.2011.02.012.

Kudra, T. 2007. "Energy Aspects in Drying." *Drying Technology* 22 (5): 917–932. doi:10.1081/DRT-120038572.

Kudra, T. 2012. "Energy Performance of Convective Dryers." *Drying Technology* 30 (11–12): 1190–1198. doi:10.1080/07373937.2012.690803.

Kumar, Ch., M.A. Karim, and M.U.H. Joardder. 2014. "Intermittent Drying of Food Products: A Critical Review." *Journal of Food Engineering* 121 (1): 48–57. doi:10.1016/j.jfoodeng.2013.08.014.

Manzetti, S., and O. Andersen. 2015. "A Review of Emission Products from Bioethanol and Its Blends with Gasoline. Background for New Guidelines for Emission Control." *Fuel* 140: 293–301. doi:10.1016/j.fuel.2014.09.101.

Markowski, A., and A.S. Mujumdar. 2006. "Safety Aspects of Industrial Dryers." In *Handbook of Industrial Drying*, edited by A.S. Mujumdar, 28. Boca Raton, FL: CRC Press.

Masum, B.M., H.H. Masjuki, M.A. Kalam, I.M. Rizwanul Fattah, S.M Palash, and M.J. Abedin. 2013. "Effect of Ethanol-Gasoline Blend on NOx Emission in SI Engine." *Renewable and Sustainable Energy Reviews* 24: 209–222. doi:10.1016/j.rser.2013.03.046.

Najafi, G., B. Ghobadian, T. Tavakoli, D. R. Buttsworth, T. F. Yusaf, and M. Faizollahnejad. 2009. "Performance and Exhaust Emissions of a Gasoline Engine with Ethanol Blended Gasoline Fuels Using Artificial Neural Network." *Applied Energy* 86 (5): 630–639. doi:10.1016/j.apenergy.2008.09.017.

Piatkowski, M., M. Taradaichenko, and I. Zbicinski. 2015. "Energy Consumption and Product Quality Interactions in Flame Spray Drying." *Drying Technology* 33 (9): 1022–1028. doi:10.1080/07373937.2014.924137.

Pisecky, J. 2012. "Components of a spray drying installation." In *Handbook of Milk Powder Manufacture*, edited by V. Westergaard and E. Refstrup, 51–85. Copenhagen: GEA Process Engineering A/S

Polanczyk, A., P. Wawrzyniak, and I. Zbicinski. 2013. "CFD Analysis of Dust Explosion Relief System in the Counter-Current Industrial Spray Drying Tower." *Drying Technology* 31 (8): 881–890. doi:10.1080/07373937.2012.736909.

Rajan, S. 1984. "Water-Ethanol-Gasoline Blends - Physical Properties, Power, and Pollution Characteristics." *Journal of Engineering for Gas Turbines and Power* 106 (4): 841–848. doi:10.1115/1.3239648.

RSBP Spol. s r.O. 2021. Accessed February 25. https://www.rsbp.cz/.

Sallevelt, J.L.H.P., A.K. Pozarlik, M. Beran, L.U. Axelsson, and G. Brem. 2014. "Bioethanol Combustion in an Industrial Gas Turbine Combustor: Simulations and Experiments." *Journal of Engineering for Gas Turbines and Power* 136 (7): 071501. doi:10.1115/1.4026529.

Shirneshan, A. 2013. "HC, CO, CO2 and NOx Emission Evaluation of a Diesel Engine Fueled with Waste Frying Oil Methyl Ester." *Procedia - Social and Behavioral Sciences* 75: 292–297. doi:10.1016/j.sbspro.2013.04.033.

Sobulska, M. 2019. "Flame Spray Drying." PhD diss., Lodz University of Technology.

Tavares, J. R., M. S. Sthel, L. S. Campos, M. V. Rocha, G. R. Lima, M. G. Da Silva, and H. Vargasa. 2011. "Evaluation of Pollutant Gases Emitted by Ethanol and Gasoline Powered Vehicles." *Procedia Environmental Sciences* 4: 51–60. doi:10.1016/j.proenv.2011.03.007.

U.S. Department of Energy National Energy Technology. 2003. *Pulse Combustor Design A DOE Assessment DOE/NETL-2003/1190*. Morgantown, WV, Pittsburgh, PA, Tulsa, OK. www.netl.doe.gov.

Wawrzyniak, P., A. Polańczyk, I. Zbicinski, M. Jaskulski, M. Podyma, and J. Rabaeva. 2012. "Modeling of Dust Explosion in the Industrial Spray Dryer." *Drying Technology* 30 (15): 1720–1729. doi:10.1080/07373937.2012.718305.

Index

acetylene 107, 109
air flow rate
 as a parameter of FSD 11, 18–20
 effect on flame temperature 42, 44, 83–84
 effect on particles flow 27, 35
 effect on particles morphology 76, 78
 effect on powder properties 81–82
 in electrode fabrication by FSP 108, 112
 in energy consumption tests 127
 in FSD scale up 59
 in product quality control 83–84
 in pulse combustion drying 93
 in safe operation 124
 in textile drying 88–89
air/fuel ratio
 as a parameter of FSD 11, 15, 16
 effect on particles morphology 69
 effect on powder properties 81, 83
 in safe operation 124
Antoine equation 49
AMD (arithmetic mean diameter)
 effect on flame temperature 40
 in product quality control 83–84
 in the dried powder 71–73, 76
 initial 21–27
 local 27–35
apparent density 69
 coffee powder 82–83
 in product quality control 84
 maltodextrin powder 78–81
arithmetic mean diameter see AMD
Arrhenius 48
aspect ratio 50
atomization pressure
 as a parameter of FSD 19–20
 effect on flame temperature 36–42, 60
 effect on initial PSD 21–24
 effect on particles flow 27–32
 effect on particles morphology 74–75
 effect on powder properties 79–82
 effect on PSD in dry powder 71
 in catalysts fabrication by FSP 105
 in energy consumption tests 126
 in FSD scale up 59
 in product quality control 83–84

bag filter 11, 92, 100, 130
batteries 102, 108, 111–112
benzene 103, 106
BET (Brunauer, Emmett, Teller)
 in catalysts fabrication by FSP 105

 in electrodes fabrication by FSP 112
 in medical materials fabrication by FSP 113, 115, 116
 in optical materials fabrication by FSP 113, 114
 in sensors fabrication by FSP 107–110
biodiesel 19, 131
bioethanol 2, 130–131
biofuels 1–2, 19, 130–131
biological drugs 113
biomass 1, 104
biomaterial 113
biomineralization 113
bone filling cement 113, 116
bone substitute 113, 116
Brunauer, Emmett, Teller see BET
bulk density
 coffee powder 82–83
 in product quality control 84
 maltodextrin powder 78–81

caramelization
 process 79
 products 57, 69
 reaction 78
CARS (coherent anti-Stokes-Raman) 8
catalysts 102–106
ceramic
 particles 69–70
 powder 42, 69–70, 131
CFD (computational fluid dynamics)
 combustion model 47–48
 in FSD scale up 59–60, 132
 model 44
 numerical mesh 50–51
 particle drying time 57–58
 particle injection points 51
 results 55
closed-loop systems 131
CO
 detection 107, 109
 emission 10, 130–131, 132
 in PCD 92
 in syngas 104
CO_2
 emission 130–131
 in combustion reaction 8, 48
 reduction 103–106
coffee
 dried powder 82–83
 raw material for drying 19, 20, 69

137

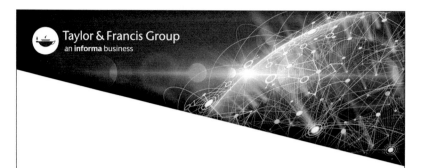